Horace William Wheelwright

Bush wanderings of a naturalist

Notes on the field sports and fauna of Australia Felix

Horace William Wheelwright

Bush wanderings of a naturalist
Notes on the field sports and fauna of Australia Felix

ISBN/EAN: 9783337196431

Printed in Europe, USA, Canada, Australia, Japan

Cover: Foto ©berggeist007 / pixelio.de

More available books at **www.hansebooks.com**

TO

ANTHONY GREEN, ESQ.

EMERALD HILL, MELBOURNE,

AS A MARK OF RESPECT FOR HIS SKILL AND TALENTS

IN HIS PROFESSION,

AND IN EVERYTHING RELATING TO THE SPORTS OF THE FIELD;

AS A TRIFLING ACKNOWLEDGMENT FOR MANY AN ACT

OF KINDNESS RECEIVED;

AND AS

A SLIGHT TOKEN OF FRIENDSHIP AND ESTEEM;

This little Work

IS DEDICATED BY HIS OLD FRIEND,

THE AUTHOR.

1860.

CONTENTS.

INTRODUCTION.

AMIDST all the wonderful revolutions that have marked the present century, no country beneath the sun has experienced such a rapid change, in so short a period, as the Australian colony of Port Phillip. Tracing the gradual development of this colony, we find that, in the year 1788, the first British convict ships landed their melancholy freight on these shores, and until the year 1803 this country remained a penal settlement. In that year the convicts were transferred to the Derwent. The first French discovery ships, under Capt. Baudin, entered Western Port bay and christened one of the large islands there, French Island, which name it now bears. Capt. Sturt appears to have been the first European visitor to the banks of the Murray in 1829. Major Mitchell, however, was the discoverer of "Australia Felix" in 1836; but Batman was the founder of the colony of Port Philip, and one of the earliest settlers in the Melbourne district. When he first camped upon the hill overlooking the now flourishing town of Melbourne, and which to this day is

called Batman's Hill, the country was in the hands of
the savage, and the kangaroo and wild dog roamed
through the surrounding bush, then as lonely as any
part of Australia.   Field sports were, of course, at that
day little heeded by the white settlers, whose sole occupation was to establish themselves in their new home ;
and the wild man, truly the monarch of all he surveyed,
held unmolested sway over hunting-grounds, then
swarming with every species of Australian game.   Tardy,
however, was the progress of advancement, and this
country might have remained in its state of primitive
wildness, had not tidings of the wonderful discovery of
the Victorian goldfields reached the Old World, and
thousands of adventurers from every clime flocked to
these shores, resolved " to do or die " in the struggle
after wealth.   Then, indeed, " a change came o'er the
spirit of the dream ;" a large and populous city sprang
up as by magic in the desert, and some little idea of the
rapid rise in value of property here may be gathered
from the fact, that, in 1853, land in the town of Melbourne sold for £210 per foot frontage, which, a few
years previous, might have been bought for £5 per acre.
The whole face of this district quickly changed.   The
woodman's axe was heard in forests which had till then
only echoed back the howl of the wild dog, or the shout
of the savage.   The country became gradually peopled.
The cockatoo settler built his log-hut on his small clearing, the wild solitude of the bush vanished before the
presence of civilized man, and the game was of course

driven back into wilder and more secluded regions by the foot of the stranger.

At the first rush to these shores, every one was far too much occupied in the search for gold, to turn his attention to the sports of the field. In fact, so all-absorbent was the thirst after instant wealth, that all regular work was for a time at a standstill. Fortunes were made and spent with a rapidity almost incredible, and it was not until hundreds usurped the place of one, that the goldfields began to lose their attractions, and men were obliged to seek a living in less exciting but steadier pursuits. Out of the thousands who yearly landed in Victoria, it was not likely that all should prosper. Many were totally unfitted for the life they had chosen; others, good men and true, but whom ill luck seemed to mark peculiarly as her own. Among these latter were men in the prime of life, brought up at home to the sports of the field from their earliest youth; and it is a matter of little surprise that when " the lecture came from the last shilling," they should turn to the gun as a means of support, and, in the freedom of the bush, unshackled by the trammels of the British Game Laws, seek an independent livelihood in pursuits which had hitherto been only an amusement: and rough and hard as is the shooter's life out here, when properly followed up, few care to leave it when they have once fairly entered upon it.

Such was my case. Six years' rambling over the forests and fells of Northern Europe had totally unfitted

me for any settled life. I had no luck in the diggings.
The town was out of the question, and to keep the wolf
from the door there were but two alternatives, to seek
work on a station, or face the bush on my own account.
I chose the latter, and never regretted that choice. I
luckily fell in with a mate in the same circumstances as
myself. The gun had often brought both of us " to
grief" in the Old World, so we agreed that for once it
should help us out in the New. Our tastes were similar.
The sphere of life in which we had both moved at home
had been the same, and therefore all those little disa-
greements and collisions which are the inevitable consequences when men of different education, training, and
tastes, are shut up together in the close companionship
of a bush tent, were avoided. For nearly four years did
we " rough it" under the same canvas, with scarcely a
single dispute, and very rarely even " a growl." We
had, it is true, at times, hardships to contend with, but
we never met troubles halfway. We took the rough
with the smooth, and whether game was plentiful or
scarce, generally had a fair share of it. Many a happy
day did we pass together in the forest. Many a good
bag of game have we brought home; and often, although
thousands of miles now separate us, do my thoughts fly
back to the old bush tent and the old comrades left
behind me ; and the chequered scenes of a wild forest
life crowd upon my mind like the " visions of yester-
day."

With the exception, perhaps, of New Zealand, where

there is scarcely a bush animal, save the half-wild pig, and no game-birds except ducks and pigeons, Australia offers less attractions to the Gordon-Cumming school of sportsmen than any foreign country; and to all who have read his diary of African slaughter, or Capt. M. Reid's *Hunter's Feast*, where American forest life is so graphically portrayed, I fear the perusal of the following pages will appear dull and devoid of interest.

I can tell of no hair-breadth escapes, no moving incidents by flood or field; and, in regard to adventure, it has been my lot, during the whole of my sporting career, to fall in with fewer than is the usual luck of travellers. But this is not my fault, and the man who writes for the amusement or instruction of his brother sportsmen, can do no more than give a true account of the sporting life of that country in which he chances to be thrown. Victoria, at the present day, occupies no mean position among the British colonies, and doubtless there are many sportsmen at home who will like to know what are the chief pursuits of the field out here, for scarcely a family in England now but has some member or friend knocking about in Australia. These are the men for whose amusement this little treatise is particularly written, and, however imperfect it may be, it has at least the truth to recommend it. I must, therefore, beg of them to take it for what it is worth.

There is no large game out here to tempt a man to wander so far for the sake of the chase alone, let his

sporting propensities be ever so keen.   One can imagine
the real sportsman, who finds his sphere too cramped and
limited at home, wishing to pay a visit to the wilds of
Africa, or the prairies of the far West.   In both these
countries the game is well worth following, and the value
of the quarry amply compensates for the risk and trouble
attendant upon its pursuit.   But in Australia the kangaroo and the wild dog are the only large animals of
chase, and the only game-birds of any size are the emu
and wild turkey.   The kangaroo, as a wild animal, stands
about on a par with the park deer at home.   The wild
turkey is now rare in the Melbourne district, and an
emu, at the present day, killed within forty miles of the
town, would be a matter of history.   But for small
game, I don't think this country can be surpassed; and
ducks, pigeon, quail, and snipe, may be killed in almost
any quantities, at the proper seasons, in those districts
where they have not been shot out.   After all, a man
can always make sure of a better day's sport here than
at home (unless he happens to be the lucky possessor of
covers and preserves of his own, and then he will most
likely stay where he is well off), without the expense of
a certificate, and with no fear of a bullying gamekeeper
before his eyes.   If he leaves the neighbourhood of the
town he can wander pretty nearly where he pleases, and
he has the satisfaction of knowing that, should all other
trades fail, he can at least get his living by his gun if he
knows how to use it, and this is more than he could do
at home.   The very absence of all those wild animals to

be found in other countries, while it renders the chase here less exciting, at least adds a greater security to bush life in Australia; and as there is now little or nothing to fear from the natives in the settled districts, the sportsman can roam the plains and forests day and night in perfect safety with no other companions but his dogs, and no requirements, in case of being benighted, except a few matches and a little salt.

There is, perhaps, no other country where a man who depends upon his gun for a living, has to work harder than he does in this. He must of necessity be camped within reach of a market for his game, which, on account of the increase of population, becomes every year more scarce and wild in the settled districts. Moreover, the shooting grounds here lie so wide apart, and so much in patches, that the shooter has to travel miles from one place to another before he reaches a likely spot, and I have many a time had to walk home six or eight miles to my tent, after sunset, with a heavy bag of game, when I was already pretty well tired out with my day's shooting. In the winter all the swamps are full, and many of the plains covered with water, and most of the best ground inaccessible except by wading. In the summer the heat is dreadful, and I can hardly say which is the most laborious, fagging on the dry plains under a burning summer's sun, or plashing through the swamps in winter up to the knees in mud and water for miles. What with the heat and the blowflies, which infest this country during the summer months, one third at least of the game is lost, and a man cannot depend at all upon the natives in the Melbourne district for assistance. There are indeed but few left, for disease and intemperance are sadly thinning their ranks, and these have become so lazy and fond of grog, since their intercourse with the white man, that they care little for work; and if they do bring you in a couple of ducks, or a bag of eels, they know their full value. The shooter here must trust entirely to his own exertions, and if he does chance to " drop on " a little lot of game, keep it to himself, for there is now as much competition in this respect as any other.

When I first commenced shooting, our great drawback was the difficulty of finding a market for the game. Melbourne was our only mart, and as we had then no horse, and no means of getting the game up without carrying it ourselves, a journey into town on foot at night, with a heavy swag of game (for the small game in the summer must be sold early in the day after it is killed), after a hard day's shooting, was no joke; now, however, there are hawkers at every fishing-station along the coast, who will buy the game at a fair price, and, as the country opens out, there will be many other places, where the shooter can make a far better living than in the vicinity of Melbourne. But let him be camped where he may, he should by all means endeavour to keep a horse, for many a weary mile's walk will this save him. An old " crawler," good enough for his work, will cost him but a trifle, and a season's keep in the bush stand him in little more than a pair of hobbles or a tether-rope.

After these few preliminary remarks I shall proceed to notice those animals and birds which form the chief pursuit of the sportsman out here. But let the reader bear in mind that I never camped more than forty miles from Melbourne in any one direction. Still, in one place or another within that district, I fell in with nearly every species of game peculiar to Victoria, and my Notes will give a pretty fair idea of the field-sports of Australia in general.

BUSH WANDERINGS

OF

A NATURALIST.

———◆◆◆———

CHAPTER I.

THE KANGAROO AND THE WALLABY.

THE *Kangaroo* (the koorah of the natives) may be called
the Australian deer, and being the only large wild animal
of chase in the country, deserves something more than
a casual notice. Of the large kangaroo I fancy we had
two distinct species in our forests, and a smaller variety
called the wallaby; of which animal, I believe, there are
several species; although the common *wallaby* is the only
one met with in the Western-port district. Altogether,
between twenty and thirty species of kangaroos exist in
Australasia.

The singular form of the kangaroo is doubtless familiar
to all who are likely to look into these pages; it is one
of the few animals whose habits are strictly terrestrial,
which, although by nature furnished with four legs, use
only the two hind ones as organs of progression. These
hind legs at first may appear disproportioned to the size
of the animal, but, upon examination, will be found beau-

B

tifully adapted to their purpose. They are three-jointed, the thigh-bone similar in shape to the shoulder-blade of other animals; being broad, and deeply grooved for the insertion of the powerful muscles which give such force to the spring of the kangaroo. The second joint is very long, being nothing more than bone and sinew, and the tendon, which runs down it, behind the hock, into the foot of the animal, is immensely powerful. The foot, which forms the third joint, is from 12 to 18 inches long, according to age, and is tipped or armed with a thick sharp-pointed nail, two to three inches long. It is also furnished with a smaller nail on the outside, higher up, and two small claws joined together inside the joint opposite; and a thick, leathery, rough kind of skin runs down behind the hock to the toe, and this is spongy on the ball of the foot. What would be the hock in another animal, appears to be the heel of the kangaroo; for you often see the print of the whole lower joint of the leg in soft ground: I fancy this is when the kangaroo is running slowly. I am no comparative anatomist, but I should say that both the outer and inner formation of the animal would form a beautiful study. The fore-arms are short, and the paws broad and large, resembling those of a beast of prey, being armed with five long sharp claws. These marsupial animals form a class to themselves, otherwise it would be difficult to assign a place to the kangaroo, were we to take the feet or teeth as a guide. In form, the hind leg is similar to that of a hare, and when in an upright position, the kangaroo rests upon its hind feet and haunches, after the manner of a squirrel; the tail stretched out at full length along the ground, not, as I have seen it represented in a picture, curled up like that of a rat, for the kangaroo cannot bend its tail. When running, it springs from the ground in an erect position, propelled by its powerful hind legs and balanced by its tail, holding its short fore-arms well into the chest, after the manner of a professional runner. Thus it bounds lightly and easily along, clearing any obstacles, such as trees, and even low fences, in its stride. I never fairly measured one of these strides or springs, but I am certain, when hard pressed, an "old man," or "flying doe," will clear nearly ten yards at a spring. The long tail materially assists them in running, and its measured thump may be heard on the ground long before the kangaroo itself appears in sight in the thick forest. It is a curious fact, that a wounded kangaroo very often breaks the hind leg in struggling; and I once knew an "old man" snap the bone just above the hock, as short as a carrot, in taking a spring. The general height of a full-grown kangaroo, when sitting upright, is, perhaps, about 5 feet. The largest, I think, that I ever killed, measured 9 feet 6 inches from the tip of the nose to the end of the tail when stretched out on the ground. The tail is very thick at the root, gradually tapering to the end; and in an old kangaroo will be 3 or 4 feet long, and weigh 10 to 15 lbs. I have, however, seen them over 20 lbs. In shape the head much resembles that of the fallow-deer; but the "lachrymal sinus," peculiar to that class of animals, is wanting. The countenance is mild and placid, but, like the sheep, we rarely see two exactly alike. The eye is bright; the nostrils not very wide; the ears large and pricked; and many of the males have a marked Roman nose, like that of an old ram. In bush parlance, the old male kangaroo is called an "old man;" the young female a "flying doe;" and the young one, till eight or ten months old, "a joey." The weight of a full-grown doe, or young buck, just killed, will vary up to about 120 lbs. Some of the "old men" reach to an immense size, and I have often killed them over 2 cwt. A hind-quarter and tail, the only part sent to market, of a young buck, or flying doe, will average about 50 lbs. when skinned and dressed. There is a good deal of flesh on the hams and back, but a great proportion of bone. The tail makes a very rich soup. The fore-quarter is very light, the chest deep, and there is something peculiar in the shape of the ribs. The kangaroo is a good swimmer, and when hard pressed will take to the water as readily as a deer. Mr. Gould mentions a kangaroo which swam for two miles through the sea, one mile being against a sharp wind and heavy waves.

The kangaroos vary much in colour, according to age and sex. The general colour, however, is dark mouse-brown on the back, lighter on the belly and flanks. The wool or hair is very fine, soft, and close; and I have seen a strand on the back of a winter skin nearly 2 inches long. Not that I fancy the wool itself could be ever used for any domestic purpose; but I think that the skins, when properly dressed, would make famous linings for a winter cloak, or pells in northern climes. They make an excellent apron for a gig or dog-cart; and when lined, are very showy. In the rutting season they have a very red tinge underneath, and we occasionally see one with the whole body-colour approaching to light chestnut. As they advance in age, the colour appears to fade; and I recollect a pure white " old man " in one of the mobs at Western-port. He suddenly disappeared, but I don't think any one shot him, or we should have heard of it. I do not, however, mean to infer that all the old kangaroos would become white if they lived long enough. I consider a pure white kangaroo nothing more than a very rare Albino variety. We occasionally meet with such anomalies among birds; and I recollect a milk-white teal, which flew with a mob of black-duck on Langhome Swamp out here. The skins are, of course, best for fur in the winter. When the skins are well picked, and properly dressed and sewed, a kangaroo rug beats any other for bush-work (but for out-door use they should be tanned); and a pair of kangaroo-leather boots are a real luxury to any old gentleman whose feet are tender, and who wishes to preserve a favourite corn.

In stretching the skins, the shooter should try to get them as square as possible; and this is best done by stretching them on sticks, something after the fashion of a boy's paper kite. But they will dry just as well if stretched and nailed out against a tree, and the tanners

think as much of them if they are merely thrown over a
pole to dry. In skinning a kangaroo, get the neck-skin
as full as possible; and this depends much upon how you
open down the fore leg. If the skin is to be dressed with
the hair on, and the head and feet perfect, as a curiosity,
be careful in skinning out the toes, and cut as much flesh
from the lips and ears as possible, and soak the skin in a
strong solution of alum and saltpetre, or the feet and ears
will go. Be careful that no blood clots on the skin, or the
hair will very likely come off and leave a bare place when
dressed. If skinned and dried properly, it may be sent
to the tanners or curriers at any time; and although the
shooter can prepare the skins himself, the process is long
and tedious, and if he wants to make a rug, he had best
have the skins dressed by a currier.

We had, I fancy, two distinct species of kangaroo in
our forests. The large one, which we used commonly to
kill, and this we found in large mobs both in the timber
and on the plains, and a rather smaller variety, darker in
colour and redder under the belly. These were generally
in more secluded situations, among the honeysuckle scrub
in deep gullies, in smaller droves, rarely exceeding a
dozen.

The flesh of the kangaroo is very inferior to venison in
flavour, and in juice and nourishment not to be compared to mutton. It tastes dry and insipid when dressed
bush fashion, but the tails make famous soup when
served up by Mr. Williams, in Melbourne, as " kangaroo
steamer." There is rarely any fat inside the carcass,

and it is a curious fact that dogs never appear to thrive on kangaroo, especially if they eat it raw, although they soon get fat on opossum. We used to make a good soup of the heads whenever we could get vegetables, which are not always at hand in the bush. My general mode of bush cooking was simple and expeditious. Just cut off a steak from any fleshy part, and throw it on the ashes to grill; and I always fancied a kangaroo-steak or even a bird, used to taste better and more juicy when dressed this " lazy bed " fashion, than in any other way. Some persons seem to think that there is no nourishment at all in the meat, and there is a great prejudice in the bush against it. This is a " vulgar error." My old mate and myself lived upon it when in the forest, and I know we did our work as well as any two shooters. Perhaps at times we might have preferred a beef-steak; but as we got the kangaroo for nothing, we just used it and made no invidious comparisons. " Spare the damper but pitch into the kangaroo, lads!" used to be our bush motto when flour was scarce. The young bucks and flying does are the best eating; the old men are tough and stringy. There is an immense deal of blood inside the kangaroo, and the flesh, unlike that of any domestic animal, does not appear to be worse from being hard-driven just before death. The meat is dark in colour, soon dries, and in appearance and taste is similar to poor doe venison.

In habits the kangaroo much resembles both the sheep and the fallow-deer. Timid and shy, their senses of sight, hearing, and smell are most acute. Like the hare, they appear unable to see an object directly in front of them when running, at least I have often stood still and shot one down as it came running straight up to me in the open forest. It is not a ruminating animal, and the four long front teeth, two in each jaw, are sharp, flat, and double-edged, peculiarly adapted for cutting or browsing; and the thick blunt crushing molars betoken a purely herbivorous animal. They are very gregarious, and are always to be met with in smaller or larger droves. I have often seen as many as a hundred and fifty in a drove, and our general mobs used to average fifty or sixty. After the rutting season, the old men will often draw away from the mobs and retire by themselves to the thickest scrub. Each drove frequents a certain district, has its particular camping and feeding grounds. The mobs do not appear to mix, and when the shooter once obtains a knowledge of the country, he has no difficulty in planting himself for a shot. Their camping-grounds are generally on some open timbered rise, and they have well-trodden runs from one ground to another. They feed early in the morning and at twilight, and I think also much by night. I fancy we might have shot them at night by a fire of dry wood lighted in a long-handled frying-pan, after the manner of torch-shooting in America; and this plan would also succeed with opossums on a dark night. But the difficulty would be to find the right kind of wood out here, for I know of no resinous trees in these forests. A good bull's-eye lantern might perhaps answer. The kangaroo lies up by day during the hot summer weather, in damp thickly-scrubbed gullies, in the winter on dry sandy rises. Here, unless disturbed, they will remain quiet for hours; and it is a pretty sight to watch a mob camped up, some of them playing with each other, some quietly nibbling the young shrubs and grass, or basking in the sun half asleep on their sides. About Christmas the young ones appear to leave their mothers' sides, and congregate in mobs by themselves. I have seen as many as fifty running together, and very pretty they looked. The kangaroo is a very clean animal. Both sexes seem to keep together, and, except in the rutting season, when desperate battles take place between the old males, they appear to live at all times in a state of domestic felicity. As far as I could see, the sides run pretty equal. Like sheep, they can be driven in almost any direction that suits the driver, and a good driver is half the battle in kangarooing. It is next to impossible to turn a mob of kangaroo when fairly off; they may divide; but they will keep on the way they are heading. Like sheep, they always follow a leader. Their principal food appears to be the tender sprouts of small shrubs and heather, quite as much as grass; but there is a small kind of spike-grass, brown on the underside, called the kangaroo-grass, to which they are very partial. They will also come at night into the small bush inclosures, and nibble off the young blades of wheat, oats, &c. I often fancied they might be kept out of such places by encircling the

fence with "sewells," which we used when deer-shooting
in the forests at home. These "sewells" are long lines
of packthread, with two white feathers tied crosswise on
the line, about a yard apart, strung up a yard or four feet
from the ground on sticks. I never knew a fallow-deer
face them. I think we might have used them with
good success in driving kangaroo; but until the game
becomes scarce and more valuable, the hunter will rarely
go out of the old-fashioned routine to procure it. Although the kangaroos feed off the ground, they do not
always appear to use the fore paws as a support, but
crouch down. I have only now and then observed
them browsing off the trees in a standing position, and
I wonder we do not oftener see them feeding in this
manner, for which their upright posture and fore-arms
seem peculiarly adapted. When in confinement, they
will eat bread, of which they seem very fond, holding it
in their fore paws, and nibbling it like a squirrel. They
are very subject, in the bush, to tape-worms, and I
have taken dozens out of the stomach of one which I
have been cutting open. Like the sheep, they can go a
long time without water, and I never could detect them
frequenting any particular water-holes at night for the
purpose of drinking. I have known their camping-places
on some of the plains miles away from any water-hole.
They appear to keep much in the neighbourhood of cattle.
The kangaroo is altogether a very domestic, interesting,
inoffensive animal, and I often regretted that we had no
better or wilder substitute for the red deer in this country.

As most of my readers probably are aware, the kangaroo, like nearly every other animal indigenous to Australia, is "marsupial," *i. e.*, the female is provided with a pouch outside the bottom of the stomach, in which are the teats, to one of which the young fœtus is attached during the period of gestation, I believe about sixty days; and when fully formed,—as soon in fact as the young one begins to live, it becomes detached from the teat, which now supplies it with milk. When the young one leaves the teat, it is in an equal state of development to the new-born offspring of any other animal; in fact, this pouch appears to be the womb of all these marsupial animals, and not, as many suppose, merely a place of refuge in which the old mother carries her young. Here the young one at first principally lives, till able to run at the foot of the mother; but even then, when danger is near, it tumbles head over heels into the pouch for protection; and it is wonderful how quickly the old doe can pick up the joey when running at full speed, and shove it into the pouch, its pretty little face always outside. There she carries it till hard pressed, when the love of life overcomes the love of the mother, and she then casts it away to save herself. This, in bush phraseology, is termed "dinging the joey." I once saw an eagle-hawk chasing a doe kangaroo with a heavy joey in the pouch through the forest. The cunning bird kept stroke for stroke with the kangaroo, which it hardly dare attack; but it well knew, as soon as the old mother became exhausted, she would cast away the young one.

Two ounces of kangaroo-shot from my gun, however, stopped the eagle's gallop: I might have killed the old kangaroo as well, but had not the heart, after seeing the struggle she was making to save the life of her offspring.

It is a curious fact that these marsupial animals should be exclusively confined to Australia, a country which, as regards zoology and botany, stands lower in the scale of creation, according to geologists, than any other. It is true that the racoon and opossum of America belong to this class; but they are only marsupial in a mean degree. The breeding habits of these animals were long a matter of doubt, and for a while it was hard to determine in what class they should be included. The peculiar formation of the generative organs was in itself enough to puzzle the best anatomists, nor was it likely that much light could be thrown on the subject by those men who, till lately, had the only opportunities of studying their habits in a state of nature. Much has been cleared up, but much still remains hidden in mystery. What always puzzled me the most was (knowing as I did their breeding habits) how close the young fœtus become attached to, or, more properly speaking, grow on a teat, in a pouch *outside* the stomach. I have killed old doe kangaroos at all seasons, and always during the period of gestation have found the young one tightly glued on to the teat by the mouth, even when scarcely so long as my finger. But I never by any chance found the rudiments of a young one *inside* the stomach.

By some naturalists these marsupial animals are placed

very low in the scale of the mammalia; and, judging from their anatomical structure alone, this classification is probably correct. But some go so far as to contend that it is warranted by the deficiency of intelligence exhibited both in their habits and physiognomy. To this reasoning I decidedly object. The brain is not so fully developed as in the true mammal, and their anatomical structure and inward functions show an affinity to the "oviparous vertebrata;" in fact, they appear to form a sort of link in the chain of creation between a higher and lower class of animal. But, however peculiar and imperfect their formation may be, when compared to the higher mammalia, in every other respect they stand fully on a par. There is nothing monstrous or ill-shaped in their outward appearance, and I am sure that the countenances of all, especially of the opossum and little native cat, are peculiarly intelligent. The habits of all are well adapted to their mode of life; and any one who has had opportunities of watching them in a state of nature, will agree with me that no deficiency of instinct or intelligence is exhibited in any of their actions.

The kangaroo I consider particularly gifted with all the attributes of a wild animal, fully equal to the deer in the senses of sight, hearing, and smell; and although the mode of progression may be different, in a mile race I should certainly "stand on" the kangaroo. It is my humble opinion that, as far as regards intelligence and instinct, the kangaroo, opossum, and native cat of Australia, stand quite as high in the scale of creation as the deer, hare, or ferret, of the old world; and with regard to those most singular animals, the duck-billed platypus and spiny ant-eater, although their incongruities of form are much more apparent, we shall find the formation of both well adapted to the habits of the two animals, the one of which passes the greater portion of its life under ground, the other under water; and I cannot see that they are in any degree inferior to the mole or hedgehog of Europe.

Geologists contend that these marsupial animals existed, in all probability, in the earlier periods, while the great work of creation was in progress, and before the true mammalia appeared on the stage; that they are, in fact, but an imperfect type of that class. Be this as it may, one thing is clear, that they alone have preserved their true character through ages, during which the earlier inhabitants of our earth have been swept away; and, unlike them, they appear to have been but little affected by the revolutions and changes which time has wrought in the aspect of our globe. And this very fact would, in my opinion, warrant us in assigning them a higher place than they now occupy.

When the joeys are strong runners, there is always a ready sale for them in town, as pets, at £1 each. They come into good season early in October, and may be caught up to Christmas. The best way to get them is to pick the old does out of the mob as they come lumbering up with the joey in the pouch. As soon as tho mother is shot down, run up, secure the young one, tie its hind legs, and put it into a bag, with a hole for the head to come out; give it a little milk and flour mixed, at intervals of every four hours, out of a bottle, like a cade lamb, for the first few days. Warmth is their principal requirement at first, and it is a good plan to put a weak joey into a bag lined with opossum-skins. Lay them out on the grass for a few hours in the middle of each day, and when they begin to nibble grass, they are fit for sale, or they may then be tethered out by day, near the tent, with a chain and collar. It is wonderful how fearless they are, and how soon they become tame. They make sweet little pets till they become too large; they are very fond of warmth at all times, and when allowed to run loose, soon take charge of the whole place. They have a shrill chatter like a monkey when angry or frightened, and at times are very spiteful.

The old doe kangaroo has but one young one at a birth, although there are three teats in the pouch, and they only breed once in the year; but they certainly are very irregular in their seasons, for I have killed a strong running joey in July, and once shot an old doe in December with a small joey on the teat. The general pairing season, however, appears to be about January or February, and the joeys, strong runners, in September. If ever the kangaroo is deemed worth preserving here, none should be killed from October until the following March, and even then the heavy does should be spared as much as possible. It would seem better, in the eyes of a sportsman at home, not to begin killing them till the

joeys could run, and to leave off just before the pairing
season; and in the old world this would be the proper
arrangement. But here the autumn, instead of the
spring, is the breeding season, and the winter the only
time when any profit is attached to the chase of the
kangaroo. The regular shooter will of course give them
a rest after October; for he has then the small game to
occupy him. The carcasses are worth nothing for the
market in the hot weather, and one winter's skin is
better than two summer ones.

Although, during the two seasons that I was kangarooing, we spared neither age nor sex, but killed them
at all times, I should have felt far better pleased, and
shot with much more satisfaction, if I could have known
that I was killing my game at fair and proper seasons.
At present, the kangaroos appear to be regarded as
nuisances in the bush, and every means are used to
exterminate the race: they are snared, shot, and run
down with hounds, just for the sake of killing them, and
the carcasses left to rot in the forest. This does, indeed,
seem a shameful waste of one of the bounties of nature.
We scarcely ever now see a kangaroo within thirty
miles of Melbourne, and they will soon become scarce
even in the wilder country. I am hardly competent
to judge how much damage they do to the grass in
the wild districts. Of course, among cultivation, they
must be far worse than the hare or rabbit at home, and
no one can blame the farmer for trying all he can to get
rid of them; but in the large plains and forests, where

there are no more signs of the plough than in the deserts of Siberia, and where the pastures are not half stocked with cattle, one would imagine there was plenty of room for a few kangaroos. I cannot but think that the value of this animal is not yet properly appreciated : the leather is acknowledged to be finer than calf-skin, and the hides run nearly as large; yet the dealers grumble to give more than 1s. 6d. per skin; and as for the meat, it is valued little more than carrion, and this in places where it can be got for nothing, while beef and mutton cost 5d. per lb. If, however, the preservation of the kangaroo becomes a question between the occupier of the land and the shooter, and it can be proved that they are injurious to the settler in the pastoral districts, I have nothing further to say on the subject; but I do not think this is the case; and it will, perhaps, be a matter of regret, at no very distant day, that the kangaroo, which, although not to be compared to the deer, is still a valuable animal, affording good meat, and, to say the least of it, forming a very pretty and interesting feature in the Australian forests, shall have become, like many animals and birds in the Old World, a theme of bygone days and a mere matter of history.

Although harmless and inoffensive when unmolested, nature has furnished the kangaroo with a dreadful weapon of defence in the powerful hind claw, with which it can rip up a dog, like the tusk of a boar; and I have seen a large kangaroo take up a powerful dog in its fore claws, bear-fashion, and try to bite it. I never but once had

one turn on me, and this was an old male which I had
knocked down, and when I went up to it on the ground,
it sprung up and came at me: he luckily fell from
exhaustion as I stepped back. Like deer, when wounded,
they will often take to water, and, if they get a dog in
their claws at such a time, always try to drown it. But
I do not believe in the fiction that they will carry a dog
to a water-hole for that purpose. They are exceedingly
tenacious of life, and, when wounded mortally, will run
for a long way, till they drop from internal hemorrhage.
As soon as they are down, the best plan is to cut the
throat; and be very cautious, in going up to a kangaroo
apparently dead, to keep out of reach of the hind foot,
for, in the death-struggle, the kicks are often very
dangerous. I have been twice knocked off my legs as
clean as if bowled down by a cricket-ball; but I was
luckily, in both instances, close to the kangaroo, and
was struck with the flat of the foot instead of the
sharp claw. I never heard them utter any sound,
except when wounded: their cry of pain is a loud hoarse
groan.

The best kangaroo-ground in Port Phillip is the Western-
port district, and begins about thirty-five miles south of
Melbourne. From hence down to the Heads is a wide
promontory, covered with deep forests intersected with
plains, about forty miles long, bounded by Port-Phillip
Bay on the one side, and Western-port Bay on the
other. Here, such is the wild nature of the country,
and so well is it adapted to the habits of the kangaroo,

that it seems as if they could never be shot out; although, of course, as the country becomes more peopled, their numbers must decrease. During the two seasons I shot here, I am certain considerably more than 2,000 kangaroos were killed by our party and another within a very limited distance, and we were camped on the very edge of the good ground nearest to Melbourne. I fancy the great breeding-grounds lie back in the wild undisturbed forests and plains between this and the Heads, and perhaps they draw down, in a kind of migration, into the more open country. I know no kangaroo-ground at all on the other side of Melbourne within the same distance. The country there is principally plains, with little or no timber. A few small flocks are met with under the Dandenong ranges, and there is a good kangaroo-ground up by the Yarra; but, according to all accounts, no country near Melbourne is equal to the Western-port district for kangaroo.

There are several methods of killing the kangaroo. Coursing them with kangaroo-hounds; snaring them; stalking them in the timber with rifles; and our old method, which is by far the best of any,—planting three or four shooters in a line through the forest, and sending a man on horseback with dogs round the kangaroos, to drive them up to the guns.

Coursing them with good hounds, in an open country, on horseback, is fair work and good sport for men who have not to get their living by the chase. It is, in fact, the aristocratic mode of kangarooing. The breed of

kangaroo-dogs in use out here, is a large broken-haired
Scotch deer-hound; the general colour red, or badger-
pied. A good dog of this kind is valuable; but we meet
with so many cross-bred mongrels, that half the dogs
which are called kangaroo-hounds are hardly worth their
keep; and I do think a lazy, half-starved, good-for-nothing
kangaroo-dog is the biggest loafer one can see about a
tent. A brace of good dogs will soon "stick up" a
kangaroo in the open, if they start on fair terms,
and in wet weather, when the ground is greasy, it is
long odds against the kangaroo. In the beginning of
winter, a three-parts grown kangaroo is easily ridden
down. It requires a little judgment in a dog to pull
down a kangaroo at full speed, and save itself from the
hind claw. An old dog, up to this work, will run stride
for stride with the kangaroo, and watching its chance,
will spring at the neck, and throw it down on its side.
A young dog generally manages to get in the way of the
claw, but a deep cut or two soon teach it caution.
The very best kangaroo-dog I ever knew, was an old imported snipe-nosed white Scotch deer-hound, such a one as
Landseer loved to draw. He was worn out; but although
he had scarcely a tooth left, could manage a kangaroo
single-handed, and his scars showed him an old warrior.
We never used kangaroo-hounds for our work; any bush-
dog is soon taught to drive, a sheep-dog as well as any
other; and a kangaroo-hound would have been little use
to us unless he would "show," *i. e.* lead us up to the
dead kangaroo after he had killed it; and such a dog is

scarce and valuable. For driving, a slow hound is better than a fast one.

Snaring kangaroo with a thick wire snare tied to a post or log, and set in their runs in the bush, or a paddock fence, answers well when a man is camped in a good country, and in regard to the skins, is better than shooting them. Snaring properly, however, requires no little skill and care, and an immense deal of attention. Snares set in the bush-runs are dangerous, on account of the cattle, and the kangaroo soon drops down to snares set in a fence. The snares should be visited night and morning, and a man cannot be sufficiently blamed, who sets his snares in the forest and neglects to see to them regularly; for, independent of the chance there is of a cow or dog being hung up (and I have taken more than one valuable dog out of a snare), it is an act of the greatest cruelty to let a miserable kangaroo remain for hours in a snare, struggling to free itself. I have often shot a kangaroo which must have been snared for a day or two. In the summer time here, when the water-holes are drying up, the bullocks and cows often get stuck in the mud, where they remain to die in a state of the greatest misery, unless pulled out. Sometimes when fast in, the station-master will not give himself the trouble to pull them out; and I once remember a miserable cow in a water-hole, on the plains, for ten days, which at length died there, although I told the owner of it. Had I shot it, I should probably have been blamed. I have also seen bullocks standing in a pen against a slaughterhouse, without a drop of water, or any food, for days, under a burning sun. Surely this should be prevented.

Stalking kangaroo in the forest, with a rifle, is, perhaps, the most sporting way of killing them. It has a good deal of excitement in it, and the skill of the shooter is fairly tested. We never used rifles when driving, on account of the danger of a stray ball in a mixed company. But I often used to "lurch" one on the feeding-grounds at night; in fact, I could generally reckon on a couple of shots any evening, if I went the right way to work. It requires an ounce ball, at least, for this kind of shooting; for often does the ball, especially a small one, pass right through the kangaroo without stopping it, unless it chances to hit a vital part. But stalking, except in the wildest bush, is dangerous work, if a man is not very careful; and I had one or two narrow escapes myself in our forests from rifle-balls.

I could any day kill a brace of kangaroo by walking through the thick forest, with the dogs driving them in all directions around me; and this and stalking are the only methods which a man can adopt, unless his party is strong enough for driving. The great objection to this sport, however, is that the hind-quarter must be carried home at once to the tent, perhaps two or three miles through the forest, unless near some bush road or well-known spot, where it can be left "till called for."

The approved method of preparing and carrying home a hind-quarter when killed, is to skin all the fore-quarter, and cut it away at the rib next below the kidneys, leaving them on the hind-quarter, to which the whole skin is left attached. Cut a hole through the skin at the neck, shove the tail through it, and drawing the skin up to the root of the tail, it will cover the belly of the hind-quarter. Hoist it up on your back, having a leg over each shoulder, the tail hanging down your back, with a leg in either hand in front; and to any one following, you have the exact appearance of an Italian boy carrying a large monkey. It is wonderful how light a heavy hind-quarter rides when properly balanced this way. The insides and fore-quarters are of no use, and are left in the forest, a prey to the wild dogs and eagle-hawks.

When a party has adopted the pursuit of kangaroos as a regular trade, there is no plan like driving. But for this work there must, at least, be two guns, a driver, and a brace of dogs. If a couple of parties join, it is best, for they need not interfere with one another. Each one has his separate tent, and the game can be divided as agreed upon. With us the driver had one share, and then every man took what he killed. The more guns there are the better is each man's chance of a shot. The dogs should not be too fast, and if they have a little music in them, all the better. In fact, if the driver has a horn and a deep-toned hound with him, it will much enliven the sport. Of course, there is one head man upon whom devolves the whole plan of the day's proceedings. The shooters start first, so that they may station themselves before the driver comes round the kangaroo. And one of the greatest secrets in driving is to give the shooters time

enough to get well planted before the kangaroos come
up. Of course the driver must know where the guns will
be stationed, and a good knowledge of the country is
indispensable to him. The shooters are planted across a
certain portion of the wood, in a line, about 150 yards
apart, each one choosing a good run, with the shelter of
a tree or bush. The best plan for the shooter is to sit at
the foot of a large tree, not to stand behind it, as I
have seen many do; and when the kangaroo are in sight,
be very careful not to stir a limb, or even to move the
gun, till they are well within shot. The driver goes
round on horseback with the dogs, and when well round
the kangaroo, he gallops on to them, and sends the mob
right up to the shooters. On they come, crashing
through the timber like a troop of cavalry, and " bang,
bang," puts every one on the *qui vive*. Sometimes the
mob breaks the line at one point, and only one man gets
a shot; but, after the first shot, they often divide, and
run right down the line, when every gun pours in
its broadside. Kangaroo-driving certainly beats deer-
shooting in one respect; for a man, who at all under-
stands it, is sure to have three or four shots in the day.
I always, if possible, like to be planted about the middle
of the line, or else sneak right away down below all the
other shooters, and never choose the first stand. It is a
good plan, if a shooter sees the whole mob breaking
the line together, to give them a shot, even if out of distance; for this will sometimes turn them down the line.
I always had two guns ready, and have sometimes brought

down four at a drive. Never, on any account, run out from your stand after a wounded kangaroo until the whole mob is past (a very common trick with a green hand); for by so doing you will, perhaps, turn all the kangaroo out of shot, and in return will, most probably, call down many a left-handed blessing from your next neighbour, who was probably just picking out a fair shot, and only waiting till it came near enough. As soon as the drive is over, the shooters meet, and each man's shot is canvassed. " What's hit is history—what's missed is mystery." I like to see the old hunter walk quietly up with one kangaroo over his shoulder, which he throws down without a remark, and turns back for a second, which he has left in the forest. Two or three may be seen struggling through the bush, pulling a heavy old man after them, while another is shouting for the driver to bring the dogs to track a wounded kangaroo, which he is certain has not gone far; to which request the driver, in general, pays very little attention, unless he knows his man. It not unfrequently happens that when the kangaroo come up in a line, the shooter gets two at a shot, and I have seen three brought down with one barrel. But the best "family shot" I ever saw, was made by my old mate. He shot right and left into a mob coming up to him, and got four old does, three of them with heavy "joeys" in the pouch; so that he bagged seven kangaroos at the two shots. It is a good plan, if the kangaroos are coming up gently, to whistle, and they will often stop in a line, and hold up their heads like seals. The dead kangaroos are now collected, drawn to some bush-road or well-marked place, laid in a heap, a piece of white paper stuck over them, to keep off the vermin, and after just one pipe, the bushman's *vade-mecum* on all occasions, the party proceed to another plant. So the day goes on, drive after drive, till evening, when the dead kangaroo, after the fore-quarters are cut away, are brought home to the tent; some on horseback, some on the hunters' shoulders. They are then skinned and dressed, the hind-quarters hung upon a gibbet, and the skins nailed out to dry. A hind-quarter will keep twice as long if skinned before it is hung up, than if hung up in the skin; and if dressed in a work-manlike fashion, of course looks all the better for the market.

Occasionally we were joined by some sporting friends from town, to whom the novelty of a few days' bush life adds double zest to the sport, and a grand battue then took place. These kangaroo battues always reminded me of the rabbit battues at home, when the keepers invite their friends for a day's rabbit-shooting in the forest. On such occasions all restraint is laid aside, every man is determined to be pleased, and the freedom of the sport is enjoyed alike by all, when all are on an equality. I can now recall to my mind's eye our head forest-ranger on the morning of such a day, in his rusty old bit of velveteen and white hat, coupling up dogs, bustling about, giving orders to the driver, laying down the plan of the day's proceedings, and greeting us with his cheery welcome, " Come, gentlemen, we must show you some sport to-day."

The evening of such a day is passed in all the free jollity of the bush. The chorus of many an old sporting song startles the magpies from their roost on the old gum-tree above the tent; anecdotes of days long past, and till now, perhaps, forgotten, while away the time, and it is not until the chairman passes the word, " Come, my lads, there's just a ' nobbler ' each, we'd better finish it, and turn in ready for the morning," that we cared to leave the camp fire. That sky must indeed be cloudy which never has one gleam of sunshine; and these little re-unions on occasions like the present, of old sporting friends, form some of the pleasantest breaks in the monotony of the shooter's forest life.

The great secret in kangaroo-shooting is never to be in a hurry; load with as much powder as your gun will stand, and never fire till the kangaroo is well within distance (I used to kill more within twenty yards than over it), and aim well at the neck. No. 2 was my favourite-sized shot. Slugs fly too wide; but for random shooting, a practice I never adopted, a few slugs mixed with the shot will bring down a kangaroo at a very long distance. The gun should be strong and heavy, and able to carry 6 drams of powder and 2 oz. of shot comfortably to the shoulder. I like Eley's green cartridge better than a bullet for kangarooing; for I have seen so many carry the ball away and drop dead in the bush, where they often lie, of no good to any one; and many a skin have I got by seeing the old crows rise off a fresh-killed carcass in the forest. In stalking kangaroo single-handed, no doubt the man who can use a rifle well, and always hit the kangaroo in the head or heart, will kill more than with a smooth bore and shot; but I am here alluding to driving, where there is no fear but that the kangaroo will come well within range. Nothing stops a kangaroo so surely as a charge of No. 2 thrown well into the neck, at about twenty yards; and I certainly did like to see a brace of kangaroo at full speed rolled over by a clean double shot. I may mention, not with the slightest desire of boasting, that no one on the kangaroo-ground killed their shots cleaner or got more kangaroo in so few shots as myself. My motive in adverting to the fact is merely to prove that my theory of kangaroo-shooting is correct. Let them come near enough, and aim well at the neck. Moreover, the longer the distance the more the shots spread; and it is easy to guess which skin is of the most value, one which the shots have entered in the neck like a ball, or one spotted all over with shot-holes like a colander. A kangaroo at full speed is by no means an easy shot, especially to a " new chum:" their peculiar jumping motion is very puzzling, and I always fancied it like shooting at a man hopping by steam. " Confound the looping beggars, I can't touch 'em at all," once observed an old deer-stalker to me (who had brought down many a red deer on his native hills), after missing three fair double shots at kangaroo in succession.

But I cannot say that I ever really fancied kangaroo-shooting much as a *sport*. There is a sameness in it when carried on month after month, which is very wearying, even if followed as an amusement; and at the present prices a man is not sufficiently remunerated for his trouble if he follows it as a trade. Moreover, there was too much of the carcass-butcher about it to please me, and driving kangaroo is certainly one of the tamest of all field-sports. When a man is hunting for his daily bread, he is justified in adopting the surest means of procuring it; the sport of the chase now becomes a business, and what would be deemed pot-hunting by the amateur, is looked upon as all fair by the professional shooter, who is perhaps guilty of many a poaching trick to obtain his game, which would be condemned in fair sporting. This, however, I thought nothing of, for I was shooting for my living and not for pleasure; but I never could reconcile to my mind the wholesale and wanton destruction of this animal which is now carried on all over the bush. Whenever I wanted a kangaroo for the body or the skin, I felt no compunction in killing it in whatever manner I best could; but I never shot one wantonly, and it certainly used to go much against the grain when I saw a kangaroo pulled down by dogs and left to rot in the bush, and old does shot with a heavy joey in the pouch, which is mercilessly torn out and its brains dashed out against a tree: with the exception of clubbing seals, this certainly did appear to be about the most barbarous

work I ever joined in. There is, it is true, some excitement in a day's kangarooing to the man who only now
and then joins in it, and the old hand often feels " his
heart in his mouth " as the mob come up to him, thump-
ing and crashing through the forest; and there is at
times a good deal of boisterous merriment in the day's
sport with a party of the right sort. But to me it always
appeared like coursing at home, slow for an hour and
dead for a minute; and although, when getting my
living by shooting, I had to take everything in its turn,
still I must say that I think I found less real sport in
kangarooing than in any other kind of shooting.

But men situated like myself must look to the profit
of the chase, not to the sport alone; and I think, on this
head, kangaroo-shooting, if rightly followed, beats any
other kind of winter shooting within the same distance
of Melbourne. Duck-shooting certainly was the most
profitable a few years ago, before the birds were shot
out round Melbourne; but now a man can hardly get his
living by ducks, unless he shoots with a punt and big
gun; and even then he must go up the bay; and for this
purpose will require a sailing-boat. When we were kangarooing, we used to sell our carcasses on the ground for
2s. 6d. each to a man who carried them up to Melbourne,
and we had the skins, which were worth about 18s. per
dozen : the hind-quarters in Melbourne are worth from
8s. to 10s. each, according to quality. Twenty-five we
considered a fair two-horse load in the winter, and these
we could easily get in four days : but we had the help of

the man who bought the kangaroo of us, and the use of
his horses for driving; without this assistance we must
have kept a horse ourselves, and had a third mate.
A good many may always be sold on the ground, and a
couple of men, if they were worth anything, ought to
kill two dozen weekly, and they can live well in the bush
for £1 per week. There are two great advantages attached
to kangarooing: the shooters get their meat for nothing,
and they "have their nights in," which the duck-shooters
do not. But if ever I were going into kangarooing again,
I would adopt a different system, and salt the hams
instead of selling the carcasses; I would try and get two
good mates, buy an old horse, tent, and rations for six
months, go up into a good country, shoot for the skins,
and cure the hams. There would be, besides, a few joeys
at £1 each, and opossum-skins always worth 5s. per
dozen; and if one of the party could skin and preserve
any rare pretty birds they fell in with, a good many
might always be sold.

The receipt for curing kangaroo hams, which I had
from a very old hand, was as follows; and what few we
made for home consumption were first rate, and ate as
much like reindeer hams as anything I ever tasted:—
15 lbs. of salt, 2 lbs. of treacle, 3 lbs. of coarse sugar,
3 oz. of saltpetre, ¼ lb. of carbonate of soda, mixed
cold in a tub, the brine strong enough to float a potato:
don't boil the brine. The above quantity is sufficient for
fifty hams. Cut the hams nicely into shape; if the bone
is taken out, the better for soaking, but the shape of the

ham is not so well. Soak the hams in the brine for five days, occasionally turning them: when properly soaked, hang them out to dry. If they are smoked, which adds much to their flavour, a proper smoking-house should be knocked up: a tent chimney will do. But you must only use green wood, and keep damping it, so that it does not blaze. I am sure I do not know what is the best wood to use here in smoking, for the juniper-bush does not grow in these forests: we used honeysuckle for what few we smoked. A hole dug in the ground, in which a fire of honeysuckle-cones and other rubbish is lighted, built over with a cone-shaped hut of tea-tree scrub, in which the hams should smoke for three or four days, will answer the purpose. It is always as well to have a tub of brine in every bush tent on the kangaroo-ground; for the meat is much improved by lying in it for only a night, and in the summer, when meat will not keep, a slice of kangaroo ham and a little bit of bacon is no bad relish.

To dry the meat without salt, cut it into long thin slices, light a large fire, and near this erect a frame of tea-tree poles. Place the flesh upon this frame, at such a distance from the fire that it will only dry up the juice: in about twenty-four hours the strips become hard and stiff, and will keep for months. This is the American mode of drying venison or buffalo.

We could, I dare say, have sold a good many hams at a much better profit than selling the carcasses whole as we did. Curing hams and drying skins requires a great deal of attention; and the dogs about a bush tent are generally the greatest thieves in the world, and take some looking after. The hams of the old-men kangaroo rarely turn out well.

The *Wallaby* is a species of small brush kangaroo, about the size of a yearling kangaroo. The general colour is very dark brown, and the hair considerably coarser and longer than in the common kangaroo, which animal, however, it resembles exactly in shape and habits, and is, in fact, a miniature kangaroo. I never met with the wallaby on the mainland in these parts, but I believe they are common in certain places further inland: they abound, however, in the scrub on Phillip Island, in Western-port Bay. They generally keep in the thick scrub, or on its edges, are easily shot in the runs, and this sport much resembles roe-deer shooting at home. The flesh is very good eating, and the skins worth 12s. to 14s. per dozen. The wallaby is very common in Van Diemen's Land, and on certain islands in the strait. This is the common wallaby, the only one which ever I saw wild; but there is another species peculiar to some of the high ranges inland, called the rock wallaby. This is described as being a shy, solitary animal, generally seen in pairs; is rather larger than the common species, and has a slightly brush tail. In habits it resembles the chamois, frequenting the most inaccessible ranges, and living among the rocks. It is very difficult to shoot; and this sport must be something like chamois-hunting.

D

There are several varieties of kangaroo in colour, those from the north being much lighter than our kangaroo; but I cannot say how many different species are met with throughout the country.

CHAPTER II.

THE WILD DOG—THE NATIVE BEAR—THE WOMBAT.

The *Wild Dog*, warrigal, or dingo, is met with in all
the thick forests, deeply-scrubbed gullies, in belts of
timber bordering on the large plains, and in patches of
tea-tree on the plains themselves, throughout the whole
country, of course commonest in the most unfrequented
districts, and is the only large wild animal of prey at
present known in Australia. Shy and retired in its
habits, the wild dog is rarely seen by day, unless disturbed, lying up generally in thick patches of tea-tree
scrub till evening sets in, when, like the wolf and fox of
the old world, they roam abroad in search of prey. In
habits the wild dog appears to resemble the European
fox much more than the wolf. Its shape, colour, and
general appearance, is that of a fox, although much thicker
and larger, and the colour is generally brighter red; but
the pricked ears, sharp nose, bright eye, and thick brush,
all strongly remind us of " old reynard." It is, however,
taller and heavier, and altogether a much bolder and
finer-looking animal. The colour is usually light red,
but there is a beautiful variety nearly black, which is,
however, rare, and, like the black fox of northern Europe,
only occasionally found in a litter of red cubs. The **cry**

of the wild dog at night is a long dismal howl, very much resembling the horrid cry of the Swedish wolf, echoing through the forests, making " night hideous ;" and sometimes a small pack would come sweeping by our campfire at night after kangaroo, and the chorus was then very fine, when all else was still. The wild bitch brings forth from four to six cubs, like the domestic bitch, generally in a large hollow log or old tree-root. Unlike the wolf, they rarely hunt in large packs, and if, by chance, four or five are seen together, I fancy it is an old bitch and her cubs : I have, however, heard stock-riders say that they have sometimes seen a large drove congregated over a dead carcass on the plains up country. They appeared to be much more common in our forests during the winter than in the summer, and this is also the case with the northern wolf: we had no lack of them on the kangaroo-ground, attracted, doubtless, by the carcasses that strewed the forests; and if ever we left a dead kangaroo out at night, it was pretty sure to be half eaten by morning. I believe the wild dog was never known to attack man, nor will they molest horned cattle, unless it be a cow in the act of parturition, when they will sometimes eat away the calf. Their chief food appears to be kangaroo, sheep, all bush animals, and offal, and birds; and when kept on the chain, they are " death upon " any fowls which come within their reach. They are a fearful scourge to the settlers on the large sheep-runs up the country; for, strictly as the fold may be guarded at night, a wild dog or two will occasionally creep in, and kill and maim many of the sheep; for, like the common dog which takes to worrying sheep, they will bite and tear perhaps a dozen to every one that they kill; and this is not the worst; for the sheep will often break fold, and, frightened to death, scattering themselves over the bush, may not be recovered again for days. There is a kind of venom attached to the bite of the wild dog; for the wound always festers, and sometimes mortification takes place: the bush remedy is to rub a little salt into the bitten part. Like the Ishmaelite of old, every man's hand is against them; they are shot, snared, and run down by kangaroo-dogs, whenever they can be met with; but the most certain way of getting rid of them is by poison. Take a small piece of meat, cut a slit in it, and insert as much strychnine as will cover the end of the blade of a penknife; hang it up by a string to a twig about a foot or eighteen inches from the ground. The dog never goes far to die after taking this bait; but they will carry arsenic a long way. They are difficult to shoot, being very wary; and there is no regular method of hunting them carried on here: what are killed, are shot, worried by bush-dogs, or poisoned.

The wild dog will often breed with the tame bush-dog, and the cross is generally larger and savager than the original breed. I recollect one morning about daylight going out of my tent and seeing a wild bitch with all our dogs playing round her. She made off into the forest when she saw me. One of our dogs followed her, and came back after three days, bitten all to pieces. The

wild dogs are cowardly by nature, but when brought
to bay, they make a hard fight of it, and it will give
a good bush-dog all his work to do to kill one single-
handed: they snap like a wolf. When the distemper
raged so fearfully a few years ago among the domestic
dogs out here, it extended also to the wild dogs, and
scores were found dead in the bush.

Although called the untamable dog of New South
Wales, I have seen them to all appearance as tame as
the domestic dog, and I knew a shepherd who had one
which followed him about like a sheep-dog. But
they are never to be trusted, nor do I fancy that they
can ever be made of any use to man, either for guarding
or any other purpose. The only bark I ever heard one
utter, was a kind of "yap yap," after a long howl.

The *Koala* or *Native Bear* of Australia is also a pouched
animal, and from its sluggish habits is sometimes called
the Australian sloth, about the size of a large poodle dog,
of a light gray colour, with white throat and rump, and
no tail; and a very comical-looking fellow he is, with his
round bald face, small black eyes, and square fringed ears.
The skin is very thick, and tans to an excellent leather;
the fur short and close. The legs are very powerful, and
the claws long and sharp. It is lazy and sluggish, but an
inoffensive animal, subsisting principally upon green
leaves, and is purely herbivorous. It lives in hollow trees,
and is not strictly nocturnal in its habits, for we often
killed them by day. I generally found them most common about the end of autumn, and used chiefly to see

them in the evening crawling about the top branches
of the large gum-trees, often with a young one perched
upon the rump. The habits of very few of the animals
here are diurnal, and we meet none in the Australian
forests by day (except it has been disturbed from its lair),
with the exception of the kangaroo or an old bear. The
bear must be considered as representing the monkey,
of which animal we have none here; a circumstance I
rather wondered at, considering the wooded nature of
the country and the fine climate. The bear makes a
poor figure on the ground, but will soon get up to the
top of the highest tree. They are extremely difficult to
shoot, on account of the thick hide; and it is cruelty to
shoot at them with shot, if they are any height up a
tree; but a bullet brings them down "by the run." The
flesh is eatable—not unlike that of the northern bear
in taste. It is considered a delicacy by the blacks. I
always found the bear singly. They have a loud hoarse
groan or cry, which they utter when frightened or
wounded.

The *Wombat* is analogous to the badger, and common
in most of the sandstone ranges in the country, where
they live in deep burrows, like the badger at home. It
is a thick, chubby animal, much larger than the native
bear, of a uniform brown colour, with short strong legs.
The skin is of little use as a fur, for the hair is short and
bristly. The habits of the wombat are strictly terrestrial,
and it is rarely seen by day. The flesh is eatable. It is
an inoffensive animal, living chiefly on herbs and roots.

CHAPTER III.

Two species of so-called opossum were common in our
forests: the large *Silver Opossum* and the little *Ring-tail*.
Wherever the gum or peppermint trees grow to any age or
size, there you will always find the large opossum; of
course, much more numerous in some localities than others,
and generally in the vicinity of water. The silver opossum
is something in the size and shape of a large cat, but the
tail is long, black, and brushy, the underside being
covered with black skin instead of hair. The teeth are
not carnivorous, but the front teeth are long. The toes
have long sharp claws, and it has a blunt thumb on each
hind foot. The nose is pointed, the face round, the
countenance mild, the ears large and pricked. A full-
grown opossum will weigh about 10 lbs. Unlike the
kangaroo, the opossum *can* curl its tail, and if in falling,
a dying opossum catches it round a branch, it dies in
that position, and there hangs. The skins vary much in
colour, from a dark black-brown, which species is peculiar to Van Diemen's Land, where the opossums are
larger and handsomer than in Port Phillip, to a light

silvery gray, with a reddish tinge on the belly, the common colour of our opossums; the shades, however, varying much; and we also had a variety dark reddish-brown throughout. When in full fur, the skin is very handsome, and has many rich tints. The opossum lives by day in the holes of the large gum-trees, and comes down at night to feed. Their principal food consists of green leaves, grass, vegetables, bread; and I believe they can also eat cooked meat. They are very partial to the leaves of the peppermint gum, which gives their flesh a rank taste. Their flesh, however, is eatable, for the blacks principally live upon it; and their method of cooking and eating opossum is primitive and disgusting. They throw it on to the coals, with the skin on and the entrails in, and when warmed through, tear it to pieces with their hands and teeth. There were few bush-animals and birds which we could not digest, but a tough old opossum beat us. The flesh of the little ring-tail is much more white and palatable, and if served up with rabbit-sauce, would be no bad substitute for rabbit. An opossum just warmed through on the coals is, however, the finest food for dogs. They come down to feed a little after sundown, and remain out till the laughing jackass sounds his morning call. As may be imagined, they are very destructive to bush gardens. The opossum is very nimble in its motions, and when the trees are high, is soon out of gun-shot, especially if the first shot does not bring it down. The only purpose for which we used to shoot them was for the skins, and as food for the dogs. The skins are worth about 5s. per dozen: they are in best fur during the winter; in the summer the hair is all scratched off the rump,—I could never account for the cause rightly. It is a curious fact, that the hair easily comes off a fresh-killed opossum; and when shooting for the skins, one must be very careful not to pull the opossum about till cold. I have seen the fur stripped off the whole body, just like a scalded pig. I fancy that the opossums come down from the ranges much in the autumn in a kind of migration to the low country, at least I often used to find them about the end of autumn thick in some places a few miles from the ranges, where, in the summer, they were very rare. They have a loud call, something between a scream and a chatter, which we used to hear much in the forests, especially during the pairing season. The opossum has usually but one young one at a birth; I have, however, more than once taken two from the pouch. The young one, at first, is red-coloured. The females breed but once in the year.

Every bush-dog has, to use a colonial phrase, "a rank down" on the opossum, and will hunt them up or find them in the trees at night, and stand barking under them till the shooter comes up. The opossum then acts very foolishly, for it will often only just run up out of reach of the dog, at which it will sit swearing, after the manner of a cat, without at all noticing the shooter below. A still night, after rain, with a moon just over the tree-tops, is the right night for 'possum-shooting. They are, however, very irregular in coming out, for in some nights you may beat the wood through and scarcely find any, while on another night you will perhaps find dozens in the same trees; but on damp nights they are sure to be out. When the moon only gives a doubtful light, they are not easy to see, especially in the thick trees; the only plan then is to get the tree well between you and the moon, and run your eye along each limb in the moon's rays. The ears of the opossum sticking up will often betray it, for they sit very still, doubled up on the branch, often in a cleft. When the night was clear and the trees bare, I never cared to have a dog with me; for let it be ever so well broken, a dog will have an occasional snap at the opossum on the ground, unless it falls stone dead; and the least blemish on the back spoils the skin. The shooter must be careful how he handles a wounded opossum. Hawker says, in his Instructions to Sportsmen: "Beware of a wounded coot, it will scratch you like a cat." I can say the same of a wounded opossum; and I have seen one fasten on a dog so tightly with its teeth as to be with difficulty shaken off. A pea-rifle is better than a gun to shoot opossums with, but be sure to take them in the head, or the bullet-hole will spoil the skin. The most I ever shot in one night was at a place called the "Banging Water-holes," near Dandenong: the trees were old and bare; the night still and clear. I killed ninety-three. This was an unusual occurrence; but a man may always with little trouble kill a dozen on any night in the forests where the opossums are at all thick.

The best rugs are those which come from Van Diemen's Land, made by the shepherds of snared skins;
and as these can be bought in Melbourne, properly
dressed and tanned, for about 50s. each, it is hardly worth
making them here for sale: still, every bushman should
make one for his own use; for of all the coverings in dry
cold weather, an opossum-skin rug is the best, as I can
well testify; for, the winter after leaving Australia I
spent in Sweden, and many a night, when the cold north
wind came howling through the pine forests, dashing the
snow and sleet against my window, the temperature of
the air many degrees below zero, I used to wrap myself
in my old opossum-rug and contrast the wild inclemency
of the northern winter with the sunny and cloudless
skies under which I secured these skins. If any blacks
are handy, it is best to get them to sew the skins, for a
black's rug beats any other. It takes about eighty skins
to make a good rug, and I have seen a hundred and
twenty used: of course, if the belly is used, much fewer
will do; but although the red colour gives the rug a rich
appearance, the skins are always thinner in this part. It
is best to tan the skins by throwing them, *when green,*
for three days, into a tub with a strong decoction of cold
boiled wattle-bark before stretching them. A good rug
should be at least eight feet by six, and when lined and
bound, it has a very rich appearance. To dry the skins,
nail them out against a tree with the fleshy side to the
sun, and they will dry in a day. As the back is the best
part, stretch the skins long but not broadwise: the more

they are stretched of course the thinner they are. In
cutting them out for a rug, try and get them as much of
a size as possible; mark out the square, and cut the skin
with a sharp knife inside, not laid on a board, or you
will cut the hair. When sewing them, use the carpet-
stitch, *i. e.* turn down the edges of the skins and sew
through them double. The blacks score the inside of
their skins with a kind of hieroglyphic, and I have seen
one marked representing a chart or map. This much
softens the skins. The proper way to prepare any skins,
such as opossum, native cat, flying squirrels, &c., for the
furrier, is to adopt the plan that we used in Sweden with
the foxes' skins. When skinning the animal, don't open
it down the belly, but make an incision across the vent
up each hind leg as far as the second joint; cut through
the legs and root of the tail, and draw the body out of
the skin, like skinning an eel; skin the head out right
down to the nose; slit down the feet to the toes, taking
out the leg-bones, and draw out the tail-bone. The skin
is now turned inside out. Cut a flat piece of wood as
broad as the body, but longer; point the nose end a
little, and thrust it into the skin down to the nose, drawing the skin smoothly and tightly over it. Nail the ends
of the skin at the tail to the board, put two cross-sticks
into the legs to stretch them, and hang it up to dry. As
soon as it is partially dry, draw out the board, turn the
skin the hair side out, and put the board in again; don't
let the skin get too dry before you turn it, or you will
have a difficulty in doing so. The reason you should

put the board in on the fur side first, is that the skin may not fasten hard on to the board, which it would do if the board was put in to the fleshy side first; but I fancy, if the board was well greased, it would not stick. A little wood-ashes rubbed on the fleshy part of the skin, assists much in drying it; and I have often found wood-ashes, sifted fine, an excellent preservative both for animals and bird skins, when no poison was at hand.

The *Ring-tail Opossum* is much smaller, scarcely half the size of the common opossum. The general colour is a plain dark brown, often with a very red tinge; the breast and belly pure white, the fur short and close, more bristly, and the skins are worth little or nothing for rugs. The tail is long and bare, like a rat's, with a white tip on the end. It is a pretty little animal, and soon becomes tame. They principally frequent thick tea-tree scrub, where they live in small colonies, building a drey like the squirrel at home. You do, however, occasionally at night find them in the gum-trees with the others, but they are nowhere so common as the large opossum. I have occasionally taken three young ones from the pouch of a ring-tail. Besides these, there are many Australian opossums, or Phalangists, as they are more rightly called.

We had two species of flying squirrel in our forests,— the *large black and white*, or *Magpie Squirrel*, or Flying Fox, and the little *Sugar Squirrel*, or "Tooan" of the natives. The magpie squirrel was rare in our district. It is principally found, I think, in the high Stringy-bark ranges, and they abound in the ranges on the Gippsland road. Strange to say, no opossums are found there.

I fancy the opossum is more partial to the peppermint and gums, and perhaps the same localities do not suit both. The large squirrel is of a dirty brown and white colour, the fur much coarser than that of the little sugar squirrel; the body itself is not very large, but I have seen them two feet long from the nose to the tip of the tail, and about a foot broad when the wings are spread out. These wings are nothing more than a fine flap of skin, which extends the whole length of the body on each side, and expands when stretched out to the toes of each foot. They certainly cannot fly, but they can float through the air for a long distance, always in a downward direction; and this is how they puzzle the dogs; for while they are barking under the tree, the squirrel floats out on the other side to the bottom of the next tree, which it soon runs up, and thus gives its enemy the slip. The cry of the big squirrel is a loud piercing scream.

The *little Sugar Squirrel* is not at all uncommon among the honeysuckle and small gums in all the forests, but is very difficult to shoot, on account of its small size and the thickness of the trees it generally frequents. It is a pretty little animal, about six inches long in the body, and the tail, which is flat and brushy, nearly the length of the body. The colour is light gray, white underneath, and the fur is beautifully soft and valuable, being a real chinchilla. They live by day in the holes of trees, and, like the opossum, come out at night to feed. The wing is about an inch and a half broad on each side. The little squirrel has four young ones at a birth, and I think breeds but once in the year. I don't know how it is with the wild dog, but I fancy none of the bush animals here have more than one litter in the year. The little squirrel is not gregarious, but generally dispersed in pairs throughout the forests. These animals are not true squirrels but belong to the Petaurists.

The *Cuscus* or *Tiger-Cat* is rather a rare animal, very like the British polecat in shape and size; and I fancy, like that animal, it lives much by the side of the creeks and swamps. It is sparingly dispersed over the thick bush, and I generally found them singly. The colour is deep chocolate-brown, irregularly spotted with white, and the tail, which is long and thin, is also spotted. It is strictly carnivorous; but the hind foot has a thumb, like the opossum. It is a shy, solitary animal, and rarely seen, although I have oftener killed them by day than night. They must be very destructive to the small game in the bush.

One of the commonest of all the bush animals is the little *Native Cat* or *Dasyure*, a pretty little animal, about the shape and size of a ferret; but the nose is sharper, the ears are large and pricked, and the tail is long and brushy, nearly the length of the body. The general colour is light sandy brown, with white spots; but there is a beautiful variety, jet black spotted with white. This, however, is rare and very local, and, unlike the black variety of the wild dog, is a distinct species; and a black-and-white spotted cat is never found among a litter of sandy ones. The native cat is a small beast of prey, very destructive to birds, especially poultry, and eggs.

They are common throughout the whole bush, living by day in hollow logs, old dead log fences, and holes in the ground, and at night they come out to feed on the ground; and the dogs, when hunting, generally run them up the small shey oaks and honeysuckles. You rarely see a wild cat up a gum-tree. They much frequent the belts of timber on the edges of the swamps; and I have often killed them on the beach by moonlight, coming down, no doubt, to look after the dead fish washed ashore. The little native cat is one of the most prolific animals in the bush, and I have often killed six young ones in a nest. It is marsupial; but, unlike the rest of these animals, does not appear to carry the young much in the pouch after they have left the teat. They are not at all shy; are very easily caught in any kind of trap baited with meat. A common figure of 4 trap is the one generally used in the bush.

The *Domestic Cat* sometimes wanders away from a station and turns bushranger; and certainly the largest cat that I ever saw in my life was a large black and white one which I killed in a honeysuckle scrub here. He must have been the very Nestor of colonial cats. I re-collect when a common cat would fetch a £5 note here. Now, however, they are at a discount. You rarely see a cat about a bush tent. I fancy a tent is hardly comfort-able enough for " pussey." Among the Laplanders, as long as they dwell in houses, the cat lives with them, but it rarely follows the wandering tribes that lead a bush life with their reindeer upon the northern fells.

E

I do not believe that there is any land rat indigenous to this country, except the bush rat; but of course the common gray Norwegian rat has found its way to Melbourne, and swarms in all the back-alleys and by-streets of that town.   The little mouse has also been implanted: both are to be met with about the towns as common as in England, but we rarely see either in the bush.

The *Flying Mouse* is certainly the most beautiful little animal in the colony ; not so large as the smallest British shrew-mouse, of a rich light brown colour above, white underneath.   It is a perfect flying squirrel in miniature, but the tail is flat and feathered.   It is rare, and very local, and, on account of its size, is seldom seen.   They sometimes come into the bush tents, and I have seen a family of.young ones taken out of a hollow tree.

Two other small bush animals, the *Kangaroo Rat* (putchook) and the *Bandicote* (boo), in these woods supplied the place of the hare and rabbit at home.   They were both excellent eating, and common throughout the whole bush.   The kangaroo rat is about the size of a three-parts-grown rabbit, but more slender, in shape like a rat ; the colour light brown, with sometimes a very red tinge ; the tail long, thick, blunt, and bare, tipped with white.   The hind legs are very long, like those of the kangaroo, and the feet are the same ; but they run on all fours.   They are pretty generally dispersed over all the forests, live in tussocks of grass on dry rises, and when the dogs bolt them, are very pretty snap-shooting up the country. They call our kangaroo rat the Paddy Mellan, and describe the real kangaroo rat as being nearly the size of a wallaby, and running on all fours. If such an animal does exist, I never saw it. We used to call a species of wallaby, or small yellow-bellied kangaroo, which is, I believe, found on Phillip Island, the Paddy Mellan.

The bandicote is a large species of bush rat, in shape and appearance resembling a very large shrew-mouse, but nearly double the size of a common English rat. We had, I fancy, two species; at least, we used to kill a large bush-rat of a dark brown colour, with very bristly hair, much resembling the animal which we called the common bandicote. This latter was, however, much the commonest, of a light brown colour, the rump striped with white crosswise; the under parts white, and the hind foot in shape like that of the kangaroo. They are generally found in hollow logs, and a bush-dog here has plenty of work in examining every dead log or fallen tree that it comes to, in the hopes of finding a bandicote or native cat. Both the bandicote and kangaroo rat have more than one young one at a birth; but, like the other bush animals, only breed once in the year. There are various smaller bush animals, such as field-mice and rats, to which I paid very little attention.

We used to kill a large species of water-rat in the creeks, and occasionally on the coast, with a dark brown body, yellow belly, and blunt tail, tipped with white, which we called the *Beaver Rat*. It is a little larger than the common water-rat, and the feet are large and flat. The skin is beautifully soft, and, I believe, valuable.

The duck-billed platypus, or water-mole, as it is called here, is found in the Yarra, the Exe, and many of the streams to the north and east of Melbourne, but I never met with it in the Western-port district. It is also common in many of the inland streams, and not rare in the Saw-mill Creek, on the Dandenong ranges. They are remarkably shy animals, and rarely seen, except at evening, when they come up to the top of the water, and look like so many black bottles floating on the surface, and sink down directly, if alarmed.

They only are found in fresh water, and I never saw them in any still detached water-holes. They may be shot by quietly watching the stream in the evening, and will take a bait, as a small piece of potato on a hook.

The singular form of the platypus must be well known to all; for the *Ornithorhynchus paradoxus* of New Holland has long ranked among the wonders of the world. I have generally seen them 1 foot to 18 inches long, and the shovel-bill 2 inches; the colour dark brown, the fur stiff and bristly, and I never saw the skins used for any other purpose than making tobacco-pouches. The tail is short, the body broad and flat, and the whole appearance of the animal betokens its mode of life. Although gregarious, I do not think they live in colonies, but each pair occupy a hole in the bank, often a long way under ground. I think they are amphibious, but I never saw them basking on the bank, and the position of their feet is not formed for walking on land. The foot is broad and webbed, the hind one turned outwards, and the male has a sharp spur on it, which is said to be poisonous : I fancy not. The eyes are scarcely perceptible, and the absence of teeth is compensated for by two horny projections at the root of the tongue, which are doubtless used by the animal in crushing the mollusca, on which it feeds.

It is certainly a singular-looking animal, and when first discovered, as its name denotes, was, I have no doubt, considered a paradox; but as science more clearly develops the hidden mysteries of nature, many a paradox, when viewed in the right light, is cleared up; and when we consider the habits of the platypus, we shall see nothing so very wonderful in its formation. The shape of the body is well adapted to the habits of an animal the greater portion of whose life is spent under water. The powerful webbed foot is scarcely more singular than that of the mole, and is used by the platypus as a propeller, in the same manner as the flat shovel-foot is used by the mole for a spade. The beak, or shovel-nose, is no more singular than the trunk of the elephant or the snout of the tapir, and peculiarly adapted for shovelling up shells, &c. from the bottom of the stream.

The beak, the web-foot, and the peculiar conformation of the collar-bone, and its habits of breeding, certainly show some affinity to the bird; but here all resemblance ends. As to the idea of its laying eggs, that has long been exploded: they are clearly mammals, for the female has teats.

Strange as the forms of all these animals appear to us, we may depend upon it that they still exist for some good reason, and we are hardly justified in regarding as monstrosities any peculiarities in the works of nature which we cannot understand.

We had a curious species of hedgehog, or ant-eater, common on all the dry sandy rises in the Western-port district,—the *Echidna* or *Spiny Ant-eater* of naturalists, about three times as large as the common European hedgehog, with sharp quills, about two inches long, a long tapering snout, similar to the beak of the platypus, but round and thin. And here, again, we see how well nature has adapted the outward form to the habits of every animal. It had the tongue of the true ant-eater, very similar to that of the woodpecker, and large burrowing feet like the mole. They live under-ground, very near the surface, and the dogs find, and quickly grub them out. I never saw one above ground except when caught. It is surprising how soon they can work their way into the ground out of sight; and when once down, it requires all the force of one man, with a spade, to prize them up.

This animal belongs to the same class as the platypus, by naturalists called monotremata, peculiar only to Australia. They stand the very lowest of all the mammalia.

I could never identify more than three species of *Bat* in our parts, and this little animal was by no means so common as I should have imagined, in a country abounding, as this does, with hollow decayed trees. Our common bat was a little larger than the large variety of British bat, and we had two smaller species. The great vampyre-bat is, I believe, met with in the Straits; but I never saw one, although I have heard of its being killed near Melbourne. The most extraordinary shot I ever made in my life was here, when I shot a bat and a large moth at a right-and-left shot.

Two other animals—the *Devil-Devil*, and the mysterious *Bunyip*—are met with at the present day in the wild swamps of Gipps' Land, according to the blacks. I need scarcely say that I never saw either. From what I can learn, there is a small species of panther, or wild cat, in Van Diemen's Land, which the blacks call the devil-devil, but it is not met with here; and as to the bunyip, I suspect it exists only in the imagination of the aborigines. Still I have heard old hands affirm, with the most extravagant oaths, such as an old hand only can invent, that they have stood face to face with the bunyip in tea-tree scrub; and they describe it as a large animal, like a polled cow, with carnivorous teeth.

On some of the islands in Western-port Bay, and along the coast, the common wild English rabbit has been turned out, and thriven well; and I believe, in many places out here, a rabbit-warren, properly looked after, would pay better than any cultivation. There is much poor hungry soil in Port Phillip, which is of little use for the plough, and less for pasture, mostly scrub and sand, but where many English esculents would grow, if planted wild, for the rabbits to feed on. The rabbits could always be sold in Melbourne for good prices. The native cats and hawks would be their worst enemies; but a small warren could be well looked after, and would, I am sure, pay.

The deer has been introduced into Van Diemen's Land, and has done well in confinement; and I fancy, if turned loose, would thrive here. I recollect one fallow-deer, which had somehow or other become loose, used to run wild at the foot of the Dandenong ranges, and has more than once been seen heading a mob of kangaroo. There was a talk of importing some fallow-deer to turn out before the hounds here, and great was the cry against it. I should much like to know the difference between hunting a wild dog or a deer as "bagmen." It is true that the wild dog is generally torn to pieces, whereas the deer, in all probability, would be saved; and that it does not break their hearts running them, is proved by the deer which are turned out, season after season, before the Queen's hounds in England.

It has been suggested that the alpaca might be introduced into this country from South America, and turned out wild to usurp the place of the kangaroo. That they would thrive here I have no doubt, and I believe they have been already kept in confinement. But if the experiment is to be tried on a large scale, I do not think it would answer at first to turn such a valuable animal loose as *feræ naturæ*. As long as they were kept in paddocks, and looked after like sheep, they would be private property; but if once they were turned loose, they would be anybody's game; and I do not see how they could be preserved sufficiently to allow them to gain head in the country. Nor do I fancy, wild as Port Phillip may be in some parts, it is, anywhere in the settled districts, so inaccessible as the native home of the alpaca.

There is no particular wild breed of cattle, horses, or sheep, indigenous to Australia. In fact, it would appear that this immense island had been left a barren waste upon the face of the globe, until its hidden resources should be developed by the skill and perseverance of civilized man; for so genial is its climate, and so peculiar its soil, that almost any animal or plant will thrive here, no matter from what part of the world it is imported. And this very fact, now clearly proved, goes far to refute the argument that Australia is a country fitted by nature only as a residence for the lowest class of animals, the marsupial. Whether or not, as has been hinted by a modern author, this land is as yet only in a primitive era, and may still be subjected to those changes which the study of geology proves to have taken place in the old world, must, of course, remain a vague hypothesis. In some parts of the country, up the Bass River for instance, large mobs of cattle breed in the bush, roam the forests, without a brand, as wild as any on the plains of the Brazils, or the South-American Andes. I never cared to meet what they call the tame cattle here in the bush, notwithstanding even the stockman's guarantee, "Oh, they won't hurt you." And this reminds me of a very unpleasant situation in which I was once placed when going over Sir M. W. Ridley's kennels at Blagdon, Northumberland. Of course I was accompanied by old Fenwick Hunnum, the feeder. While we were looking at the bitch pack, a quarrel broke out in the dog-kennel, and the old boy slipped out to quell the riot, quietly observing, as he shut the door behind him, leaving me alone with the bitches—"They won't hurt you, I expect." They certainly did not hurt me; but the way in which they came sniffing round me with their bristles up, one every now and then uttering a low growl, was anything but pleasant; and I was glad enough when the old man came back and exultingly remarked, with a grin on his old foxy face, "I told you they would not hurt you."

Many of these so-called tame cattle are dangerous, especially the cows, which calve in the forest, plant their calves, and go a little distance off to feed, and old working bullocks: I believe here the bulls are the most harmless. I did not so much mind them in the timber, for a man has a chance of getting behind or up a tree (I was once stuck up a whole night in a honeysuckle). But I always looked out on the plains, and whenever I saw a bullock stand sulking by itself, I always gave it a wide berth; such a one is generally "a Roosian." Of course, with a gun a man has not so much to fear, but a charge of shot will often not stop a rushing bullock. One would not like to shoot a bullock on a run; but better kill him than he kill you, and I always had a bullet in my pocket ready to slip in in case of need. I always found a good large dog the best protection.

It appears that the first convict-ships which came to these shores, in 1788, brought out with them one bull, one bull calf, four cows, one stallion, three mares, and three foals; and from these have sprung the immense mobs of cattle and horses which now wander over the forests and plains of Victoria. According to the "British Farmer's Magazine," it seems that in 1851 there were 390,000 horned cattle, and 16,500 horses in Victoria; and the sheep in this colony, which, in 1838 numbered three millions, had increased to five millions. It is most probable that the sheep were introduced into Victoria from the older colonies. In 1788 the first sheep were imported into Sydney from India; the number originally brought in was twenty-nine. These, in 1803, had increased to 10,000, and in 1846 to nearly seven millions. In 1807, the export of wool from Australia to England was but 245 lbs., and in 1855 it had reached forty millions of pounds; which, coupled with its annual ten tons of gold, ought to render this country one of the richest in the world.

I have, I believe, above noticed all the *common* bush animals of Victoria. In the wilder parts, some other species are, no doubt, met with; but these are all I know. It will be seen that there is very poor encouragement for the fur-hunter out here; but at the same time there is not a single wild animal in the forests which the bushman need fear. We will now turn to the feathered game list, which we shall find richer both in species and individuals.

CHAPTER IV.

THE EMU—THE WILD TURKEY—THE LOWAN—THE NATIVE PHEASANT.

THE *Emu*, or as the natives call it, "Ourer," is also called the Australian cassowary and is the largest bird in the colony, but is now rarely met with in the settled districts, and I can say nothing of its habits from my own personal observation. It is by no means rare in many parts of the country; but we must now look for the emu far back in the wild plains and extensive sheep-runs up country, which are rarely intruded upon by the presence of man, except it be a solitary shepherd or stock-rider. A small flock used to frequent the wild country round the kangaroo-ground, and during my stay there two were killed. They were not so very shy, and often came within range of the wood-splitters' tents. In habits and appearance the emu much resembles the ostrich; but it is not nearly so large, and wants the fine tail and wing feathers peculiar to that bird. The general colour is brownish-black, the feathers long, and clothed with fibres like hair; they can't fly, and are generally ridden down with kangaroo dogs. It is a very fat bird, and when boiled down, emu oil, like the shark's oil among fishermen, is the bushman's universal remedy for rheumatism and other bush com-

plaints. When properly dressed, the skin makes a fine
rug, which is very warm, and moreover, a bit of a " curio."
A full-grown emu will stand above six feet, and I know
that it takes two men to lift one on to a horse. The
breeding habits of the emu differ from the ostrich ; at
least, I once found two eggs, and they were not fresh,
near Arthur's Seat, on the coast here. They lay open in
a little hole in the ground, scraped among a heap of moss
and rubbish in the forest. They were rather larger than
a swan's egg, of a greenish-black colour.

The *Wild Turkey* (gollopachin) is certainly entitled to
the first place in the list of Australian game birds. It is
a species of bustard, smaller, however, than the European
bustard, and the male wants the moustache peculiar
to that bird. The legs are not so long in proportion, it
flies much better, and when in the air, rather resembles
the common turkey. It is of a light gray colour,
mottled and pencilled with black. An old male will
weigh about 20 lbs., a female 9 to 12. They generally
frequent the plains and open moors, are partial to old
sheep-folding grounds, and I have seen as many as
twenty-seven feeding together on the wide open country
towards Gelong. It is a very shy bird, and few are met
with now in the neighbourhood of Melbourne, but they
abound on the large sheep-stations up country ; they generally came into our district as stragglers, but an odd
couple or so bred in the heather ; for I have often raised a
single bird in the summer on an open moor ; and there
were certain places where we could generally see three

or four feeding, about the autumn; probably birds bred in the neighbourhood. What few I have seen killed were chance birds. It is next to impossible to get up to them on foot in the bare plains; but like all other bush-game, they take little notice of a bullock-dray or horse, and are easily stalked under shelter of these. They generally fly low, and as they rarely alter their course, the best plan, if the shooter sees one flying up to him on the plains, is to stand still, and he will probably get a shot. The wild turkey is a fine-eating bird, and worth about £1 in Melbourne, but you rarely see one in the market; for where they do abound, nobody cares to shoot for profit, and what are killed find their way to the head station. They were very rare in the Western-port district: the country is too deeply timbered. The large open plains on the sheep-stations in the interior are the peculiar home of the wild turkey.

The *Lowan*, or native hen, is peculiar to the country in the vicinity of the " Mallee Scrub," in the interior, a species of dwarf gum, about 12 feet high, and smaller scrub, so tightly interlaced with the tendrils of the native vine, as to render it impenetrable. The lowan is a plain dull-coloured bird, brownish black, a little larger than the common fowl, and lays an immense egg for its size, in the sand. The birds lay a number of eggs together, heaped up in the form of a pyramid; whence their name of the mound-building bird of Australia. They are covered and hatched in the sand, and, strange to say, the young birds are not seen till they are pretty well feathered ; I never met with this bird in a wild state.

The *Native Pheasant* is the Lyre-bird of naturalists, and the " bulla-bulla " bird of the natives, from its call-note, and is by no means rare in the peculiar localities which they frequent,—the most secluded gullies in certain high ranges. They were common in some of the gullies on the Dandenong ranges, up the Plenty ranges, at the head of the Yarra, and up the Bass River, on the eastern coast of Western-port Bay; but I never heard of one being killed on the west side of that bay. There is nothing handsome in the general plumage of the native pheasant ;—it is about as large as the pheasant at home, the body dull-coloured brown; but the beauty of the bird consists in the tail of the male, which is very long, the feathers clothed with fibres like those of the birds of paradise, in the form of a lyre, the two outer feathers curved outwards, like those of the black-cock at home. It is one of the shiest birds in the world, rarely seen on the wing, but keeps on the ground among the thickest scrub and fallen timber. It is a perfect mocking-bird, and the only way to shoot them is to lie still and call them. It is little use in a white man going after them without the assistance of a black. The blacks make periodical excursions up into the ranges, about September, when the birds are full-feathered, and come back laden with tails. Just as I left Melbourne, I saw the nest and egg of this bird brought down from the Plenty ranges. The nest was large and domed, the egg

uniform dark chocolate-brown coloured. I never killed
the pheasant, although I have often heard them on the
ranges, and I should not have noticed either this or the
lowan-bird, but that I fancy they make up the full list of
Australian game-birds, with those which I am about to
mention.

CHAPTER V.

I DO not believe that any country in the world is better adapted by nature as a home for the water-fowl than Australia. Dreary swamps miles in extent, lagoons of immense size, where the bulrush and reed vegetate in rank luxuriance; creeks and water-holes, completely hidden from the view by dense masses of tea-tree scrub, afford unmolested shelter and breeding-places for the birds; and a few years ago, when the sound of a gun was rarely heard in the solitude of these morasses and fens, the country around Melbourne must have literarily swarmed with wild fowl. When I first came into this country, the palmy days of the duck-shooter were in their zenith; the fowls and buyers plentiful, the shooters scarce. The year previous there was not a float or big gun in this part of the colony, and the first punt that ever floated on Melbourne Swamp, was built in Melbourne Street, where the market now stands, in the morning, launched in the afternoon, fitted up with an old musket, and the birds shot and sold in Melbourne before night. In this winter, £1,000 was cleared off Melbourne Swamp and its neighbourhood by the two men who launched this punt. The diggings were then in full swing, money was like dirt, and the birds sold at any price. The buyers were not particular. Many a brace of sea-gull have been sold for 5s. and I once knew a pair of old shags with their beaks trimmed up, sold for 15s. as "rock duck." But this did not last long. The duck-shooters of that day, like the diggers, never heeded the morrow, and not one laid up for a rainy day. As the birds became scarcer, the shooters increased, and prices fell, till at the present day duck-shooting is not worth following within fifteen miles of Melbourne. What a change has six years made in the appearance of this country. The swamps and lagoons near Sandridge are all drained or built on, and a railroad now passes over ground on which, at that day, four or five couple of ducks might be killed with ease in a night's flight-shooting.

Eight species of wild duck are more or less common in this district, and I believe these are nearly all the ducks indigenous to Victoria: the Mountain Duck, the Black Duck, the Wood Duck, the Pochard or China-eye, the Whistle-wing or Pink-eye, the Shovel-bill, the Teal, and the Musk Duck. I have seen one other species in Melbourne, said to have been shot in the neighbourhood, as large as the black duck, but more resembling the British gadwall in plumage. This I believe to be only a rare and occasional visitant to these parts; although I have heard that it is common in some parts further inland.

The *Black Swan* is common throughout the winter

after the young birds can fly, on all the large swamps
and lagoons; sometimes in good-sized flocks, but generally
in small companies, which I took to be the old birds and
birds of the year. Early in summer they retire to their
breeding-haunts, and we saw very little of them again
until the swamps and water-holes filled. They appear
to breed in August and September. The nest is a large
heap of rushes, and the female lays five to seven dirty-
white eggs, not so large as those of the swan at home.
They breed a good deal on some of the large islands
in Western-port Bay, and I attribute the decrease of
swans in this neighbourhood to the quantity of eggs
that are yearly taken by the fishermen in this bay.
Swan-ponds near the Heads, is also a great place for them;
in fact, they are by no means rare in this district, and an
odd pair or so breed on most of the large swamps. The
black swan is not nearly so shy as the European hooper,
and they are by no means difficult to come up to with a
punt-gun. They are a heavy-flying bird, and don't care
to rise on the wing, if they can save themselves by
swimming.

The black swan is a graceful, elegant bird, not so large
as the hooper at home; the shape of the beak is the same,
but the cere is red, and the windpipe is not folded within
the breastbone. The colour is deep black, the pinion-
feathers white, which contrast prettily with the black
plumage of the body when the bird is in the air; the
bastard wing-feathers are prettily curled. They have
a very musical call-note when passing overhead on a

still night; and I have listened with pleasure to the soft low notes of a pair of swans answering each other, while floating on the lagoon, by the side of which I lay at flight-time. At night they always fly low. The black swan does not attain its full plumage till after the first or second moult: the young birds are light mottled-gray.

The swan is hardly worth shooting here for the market, as they only fetch 5s. each, and they are a heavy bird to carry about. The flesh of the young swan is excellent, and one roasted in a camp oven generally with us formed the duck-shooter's Sunday dinner, whenever we could get one during the season. I wonder the skins are not more highly prized for the down, which is very thick. This is the only species of swan indigenous to Australia; but I once saw the real *rara avis* out here, or white swan, flying up the bay about a quarter of a mile out to sea. Nobody believed me when I mentioned it, but I pointed it out to a friend who was with me. I can't pretend to say where it came from. One would naturally think it had escaped from some aviary; but nobody at that time kept tame swans in this neighbourhood that I know of, although a pair may now be seen in the Cremorne Gardens, Melbourne.

Two species of wild geese are met with here,—the Magpie, or Tree-goose, and the Cape Barron goose.

The *Magpie*, or *Tree-goose* (ongak), is the common wild goose in this district, and, as far as I could learn, is the only common wild goose peculiar to Port Phillip. Although met with here only in small flocks, generally I

think families, there are lakes in the interior where they
swarm. I think they remained in our district throughout
the year, although we used only to see them at uncertain
periods, and never for long together. As the name
denotes, the colour of the magpie-goose is pied, dull
black and white: it is about as large as the British
brent goose, and the tail is very square. It is a singular
bird: the beak is higher in shape, and not so broad, as in
the common goose, has a palish red rough cere, and the
upper mandible is long, and has a powerful curve or hook.
It has a large warty cere, extending over the front of the
head, which is in shape like that of a game-cock, cut out
helmet-combed. The feet are semi-palmated, and formed
for perching; the claws long and sharp. I rarely saw
them either on the ground or on the water, never, certainly, in open water, although I have raised them out of
the thick reeds and grass that choke up many of the
creeks and lagoons here. They are generally perched
high up in the tea-tree scrub, where they will sit for
hours; and a curious sight it is to see them sitting
upright, with their long necks stretched out on the
watch. They have a very loud, hoarse call-note when
alarmed, nothing like that of the common wild goose.
The greatest curiosity of this singular bird, however, is
the windpipe, which has three folds, like that of the
European hooper; but, instead of being folded within the
breastbone, it lies on the left hand, outside, bedded in the
flesh. They bred sparingly with us, for I have found
the nest in a thick tea-tree scrub; and I fancy the small

flocks that we see in the autumn are families, which had been bred in the neighbourhood, and that they do not pack and make distant migrations like the wild geese at home. Although a shy bird in the open, they are by no means difficult to creep up to in the thick tea-tree scrub, and many a pair have I killed right and left. They are capital eating, and will fetch from 12s. to 15s. per couple in the market.

The *Cape Barron Goose*, the New Holland *cereopsis* of naturalists, looks like a cross between a goose and a turkey, and is only a rare and occasional visitant to our parts. It is rather larger and heavier than the magpie-goose, of a light gray colour, spotted and chequered all over with black; and the beak and feet in shape resemble those of that bird. I never saw them here but twice,—once in a small flock, and once when two pitched with the tame geese at Mordialloc (this, I believe, they are fond of doing), and which were caught alive. They soon became tame, and used to stalk about the paddock; but they were very pugnacious with the other geese: their call-note was a deep trumpet-like sound. They very little resemble a goose when walking, but put me more in mind of the Canada goose in shape than any I know. These are the only two species met with here, and neither of them appear to be true geese.

The *Mountain Duck* is the largest and handsomest of all the ducks out here, nearly as large as the bernicle goose of Europe, and in colour resembling the male sheldrake. They are generally seen feeding on the plains,

in small companies, in the vicinity of water, and as they
are very wary, and the old drake always on the look-out,
a brace of mountain duck is no mean prize. I very rarely
saw them on the water, but they pack in some favourite
lagoons, and are not difficult to come up to with a punt.
I never saw them in the creeks, but always in the open.
The old male bird utters a peculiar hoarse guttural
warning when danger is near. They breed in our neighbourhood, I have heard, in trees; and I have taken the
young birds, but a few days old, in the damp grass on
the swamps. They rarely associate with the other species,
and fetch no more in the market than the black duck,
although nearly half as large again, for the flesh is considered coarser.

None of the Australian ducks, except the black duck
and the teal, appeared to fly in large flocks; and all the
male birds had that peculiar excrescence in the windpipe
peculiar to the British wild ducks. I fancy most of the
ducks out here breed in trees.

The common wild duck of this country is the black
duck, and, whether for its flavour at table, its wild, gamy
appearance, or the sport it affords the shooter, is certainly
equal to any duck in the world.

The *Black Duck* is of a deep black-brown colour; the
feathers edged with lighter brown, a very brilliant deep
purple speculum on each wing, the cheeks and throat
rich chestnut-red. It has a peculiar snake-like appearance about the head and neck, and, with the exception
of the spinetailed swift, is, I think, the sharpest-flying

bird in the colony. A pair of good black duck will weigh about 5 lbs., and average now 7s. in the market.

The duck season here commences in the end of January, when the old birds bring their young down to the creeks (I have shot flappers, in some seasons, early in January), and should end with September, when the swamps begin to dry up, and the birds pair off and retire to their breeding haunts. After they have bred, they keep about the creeks and water-holes in small flocks or families, till the rain fills the large swamps, when they seem to congregate and frequent the open places on the swamps and plains, where there is shallow water and good feeding-ground. There is little good to be done with a shoulder-gun out here, in the large swamps by day during the winter. The black duck lays from six to eight eggs on the ground, appears to breed much in heather, and I have taken the nest in an open moor far away from any water.

There is no better sport than flapper-shooting here, and there is no country in the world by nature better adapted for it. The creeks in the summer (and in the winter the ducks are all out on the open swamps and large lagoons) are a succession of water-holes, walled in by a thick hedge of tea-tree or reeds ten to fifteen feet high. This screen appears impenetrable, and little use would it be for any one to attempt to force his way through it; but the old hand soon finds a cattle-track, which he well knows leads to water, and, creeping cautiously down it, with his retriever at his heels, he suddenly comes upon a large, clear, open water-hole, on the margin of which a score or so of ducks are floating lazily about, or sleeping in the hot sun. One barrel on the water, one as they rise, and in dashes the retriever, and first chasing the winged birds, brings to land the killed and wounded; and perhaps two couple or more of ducks lie dead upon the bank. The reeds are generally so thick round the water-hole, that many a wounded bird, and others that fall dead at a little distance, are lost. The ducks which go away rarely fly far, but drop in another hole a little lower down the stream; and two or three good shots fill the bag. You will rarely find ducks in the brackish water-holes at the mouths of the creeks. I have re-marked two things in duck-shooting out here: whenever, by day, I saw the swallows flying over the tops of the bulrushes, and dipping down into the hole, I was almost certain I should find no ducks; and at night, whenever the frogs were silent on the lagoon, ducks were on the water.

A good retriever is indispensable to the Australian duck-shooter, for the scrub and reeds around the creeks and water-holes are so thick, and the grass on the swamps and plains is so long, that a wounded bird is lost in a few seconds. A winged black duck often dives and comes up again in thick rushes, where it sits so close that even a good dog will often pass over it. It is a safe plan, after a shot, to try round the edges of the scrub outside the hole, for a wounded duck often creeps out on land. A duck-shooter here may reckon on losing half his birds without a retriever.

Of all the field sports in this colony, I think I did like a good night's flight-shooting the best. There is a charm in this silent solitary sport which I could never find in any other. When seated well in the shade, by the side of some favourite feeeding-ground, with the moon just on the wane, all is still, save the occasional cry of some night-bird as it rises from the neighbouring swamp, or the whistle of the wings of a pair of ducks, as they pass overhead, and the croaking of hundreds of small frogs in concert, the deep clock of the bull-frog joining as it were in bass accompaniment. The slight ripple of the clear water dances in the moon's silvery rays, when all at once "whish," a splash in the water, and a sharp "quack quack, quack," warns the shooter that a black duck has pitched, and the concert of frogs is hushed in an instant. This is soon joined by others, and having risen on the water three or four times to shake their feathers, and chased each other about for a few minutes, they settle down to feed. Now is a moment of breathless suspense to the shooter; the gun is quietly raised, but the birds at first are too far, or not well packed; however, at length, he gets three or four in a line, and the heavy boom of the gun breaks the stillness of night, reverberating over the swamp with a hundred echoes. It may be that some scores of birds were feeding on the lagoon out of sight, which now rise like a clap of thunder, and the air is disturbed by the wings and the cries of the birds as they fly round the shooter's head. His quick ear can well distinguish the different birds by

their varied call-notes;—the soft musical hoop of the
black swan, the sharp loud quack of the black duck, the
hoarse croak of the mountain-duck, the snort of the
shoveller, and the shrill call of the teal, are all familiar
to him; and as he gathers up his dead birds, he hears the
ducks pitching again in various parts of the lagoon,
giving him promise of a goodly harvest by morning.
When the dead birds are collected, the pipe is lit, the
gun charged, and he quietly settles himself down in his
rushy screen for another shot.   The early part of the
evening is best for this sport; the birds leave the feeding-
grounds about midnight, often go out to sea, if the lagoon
is on the coast, and return again a little before daybreak,
when they often pack on the bank of the lagoon.   So in
punt-shooting, the evening and the morning shots are
those upon which the shooter principally depends.
Where the birds are feeding well upon ground which has
been but little disturbed, flight-shooting is the best and
surest game of any with a shoulder-gun, and there is
some little difference between flight-shooting out here
and at home, where the shooter has to sit for hours,
often in sleet and drizzly rain, his teeth chattering and
his fingers so cold that he can scarcely pull the trigger.
Here a good pea-jacket will keep the shooter warm on
the coldest night, and though I have occasionally used
gloves, I never really wanted them.   The best seasons
for flight-shooting are the autumn and early winter.   In
the months of March and April, 1858, my old mate
killed upwards of a hundred couple of birds, princi-

pally black duck, at night, with his own gun, in one small water-hole close to the coast. This is the only kind of shooting in the colony for which a man really requires water-boots. As the birds generally feed in shallow water, he fetches the dead ones out himself, and he may often have to sit for hours on a tussock of rushes, up to his knees in water. Cording's Indian-rubber water-proofs are the best I ever used for this work; they are warm, perfectly water-tight, never want dressing, and, what is best of all, never get hard, and are always easy to pull on and off. They are certainly too heavy for walking much in, but for flight-shooting, boat-fishing, or any other work where the wearer is not constantly in motion, I will back them against any boots in the world. The American gutta-percha overalls are not worth anything for work. At all other times except flight-shooting, the best dress for the Australian duck-shooter is canvas or flannel trowsers and low half-boots. The climate is so fine here, that a man may wade in the swamps with impunity at all seasons of the year, and the best clothes the shooter can wear are those which dry the quickest.

The flight-shooter usually ties a black ribbon round, or sticks a small lump of mud on, the end of the barrel of his gun, to guide his eye well on the object. My plan was better, both for flight and opossum shooting. Cut a forked piece of tea-tree, the forks about six inches long, and tie it round the end of your barrel, the muzzle protruding between the two forks, which stick up one on each side, like a pair of horns. I learnt this trick of an old poacher, who had often used it with success when nailing pheasants on the perch at home.

One season I killed a good many birds to stuffed decoy-ducks floated on a piece of board, and kept in their place with a string and a stone. We used them on the Swedish coast for eider-duck, and called them " boul-van." They answered very well in the daytime in any clear open water which the ducks used, where there was good shelter for the gun. Any ducks passing over would pitch to them; but it is wonderful how soon they discover the deception, and you must fire as soon as ever they settle on the water. A fresh-killed duck set up in a natural position, with forked sticks, in some respects answers better, as it retains the natural smell. I always made it a rule to fire directly I got three in a line, no matter how many birds were scattered upon the water; for in this sport delays are dangerous. I have often killed six or eight couple of ducks in a day to stuffed decoy-birds.

The swamp-hawks were my greatest annoyance; for many a time when I have been just getting ready for a good shot, an old swamp-hawk has come sailing over and sprung the ducks; and if ever I left my decoy-birds on the water to go to another hole to see if there were any ducks, I was sure to find one, if not more, of my stuffed birds torn to pieces by the hawks.

Eley's cartridges are dear out here,—three shillings and sixpence per dozen; but I am not sure that a duck-shooter would not gain by always using a green one in his second barrel. In duck-shooting, in all cases except at flight, when I liked a loose charge best, I used the candle cartridge, and I found them quite equal to Eley's, except that they occasionally ball the shot. As every one may not know how to make them, I will give my receipt. Procure a tin cylinder that will exactly fit into the muzzle of your gun, about three inches long, something like a candle-mould; stick a cork in the bottom end and set it on a table; put the shot in it, melt some candle-grease in a ladle, which pour on to the shot till they are covered. Let it stand to cool; take out the cork when the tallow is hard, and shove out the cartridge; wrap a piece of thin paper round, and it is ready for use. I once killed a pair of black duck stone-dead at eighty yards with a candle cartridge: this was perhaps a chance shot, but I could always reckon on my birds at fifty yards; and I know this is about fifteen yards further than I could do with a loose charge. I shot with a single pigeon-gun, No. 6 gauge, 6 drams of powder, and a two-ounce cartridge. One needs a strong gun for such a charge, and I fancy cartridges shake a gun much. I am no friend to an out-of-the-way-sized gun for shoulder-shooting. The one I used I found big enough for any purpose, and quite heavy enough to carry about and bring up to the shoulder quickly. Depend upon it, the man who sticks to one gun for every kind of shooting, will bag more game than he who is continually changing; and I believe I should have done better if I had always used a strong double. The best gun for this country is a strong dou-

ble, twelve or thirteen gauge, heavy enough to carry two
ounces of shot if wanted. The shooter out here requires
a stronger and heavier gun than at home, and a
season's wear and tear, with the charges we often put in
for kangaroo or ducks, will give the best gun a pretty
good shaking. I am quite sure that I fired more shots
in one year out here than I should have done in four
seasons at home.

In flight-shooting it is a good plan to crack off a cap
before loading, to clear the nipple, for you can't always
see well. Never keep your caps loose in the pocket,
always use a small tin box. The bottom of a bushman's
pocket is generally full of fine broken tobacco, and many a
miss-fire have I had through putting on a cap in a hurry,
without seeing first if it was clear. Another hint and I
have done. Always have your powder-flask slung round
you in duck-shooting, and don't trust to the pocket. I
have lost many a powder-flask by neglecting this caution,
in struggling through thick tea-tree scrub.

Of course, punt-shooting is the most profitable kind of
duck-shooting; but it is not every one who can use a
punt and big gun; and in this country they will cost some
money. Besides, there are not many places now within
reach of Melbourne where a punt-gun will pay. There
are now as many pop-shooters as ducks about Melbourne
Swamp, and the birds are so much disturbed in all
places, that they don't pitch anywhere in such mobs as
formerly. Above Gelong, and on some lagoons near
Ballarat, I believe two or three punts are worked, and

the shooters get a fair living. Connor, who was certainly the best punt-shooter out here, stuck to it longest
of any in our bay. He had a sailing boat, and cruised
in the bay with his float and big gun on board, and shot
on the coast and the large lagoons at the Heads; but he
knocked off at last, as I suppose it hardly paid, about
three years ago. I recollect he used to make one
trip a week, and brought back about fifty couple of fowl.
If a man is camped for any certain time near a good
lagoon, he can easily manage to knock up a float himself
with a few boards, and fit up a moderate-sized gun with
a rope-breeching (I never saw any other used here), and
the whole affair need not cost him £10.

The *Wood Duck*, take it altogether is, I think, the
handsomest little duck in the colony, hardly so large as
the pochard at home, with a head and beak exactly resembling the bernicle goose of Europe. The plumage is
silvery gray, mottled with black, with beautiful long
scapulars striped black and white; the wing-speculum
very brilliant, the breast black, the head and neck chestnut, and the male has a small crest or mane. It was by
no means a common duck, at least with us, and was
generally seen in pairs or small families in some secluded
water-hole; sometimes on the water, but more often
standing on the bank. They were by no means shy, and
easily crept up to. Bred in holes of trees, and often
perched on the gum-trees by the side of the creeks. It
appeared to be rather a local bird, and rarely associated
with the other species.

G

The *Pochard* (a better name would have been " the Widgeon "), gray-back or China-eye, as we used to call it, is a dull heavy duck, very much in plumage resembling the British widgeon, but plumper and larger, being very little less than a black duck. The eye is French white. It was not very common with us. Rather local, and sometimes seen in small flocks, but oftener in pairs. It was a shy bird, and very rarely associated with the black duck, certainly never in quantities.

The *Whistle-wing*, or Pink-eye, is the smallest and tamest, but with us the rarest, of all the Australian ducks, not larger than the water-hen at home. It is a pretty little duck, of a light silvery mottle, with a faint pink mark over each eye, and a remarkably large, broad shovel bill for the size of the bird: we usually found them in odd pairs, but I have shot on some lagoons where they came in good-sized flocks.

The *Shoveller*, or " Spoony " of the duck-shooters, is something like the shoveller at home in size, shape, and general appearance, but the plumage is not so handsome. They are chiefly found in creeks by themselves, but occasionally join a mob of black duck on the plains. It is rather a pretty duck, next in size to the black duck, and, except the teal and black duck, was the commonest of all the ducks in this district. The plumage of the male is bright chestnut mottled with black, the breast dark, the scapulars long, the speculum on the wing pale blue, and the bill broad. They seemed to be partial to particular localities, and I know one creek, called the Skeleton

Creek, above Williamstown, in which I could always find a flock. The best shot I ever made at ducks in my life was in this creek. I was beating for a snipe on the banks, with a small single gun and one ounce of No. 7 shot. I fired into a mob of spoonies which were going up the creek about fifteen yards from me. I bagged eight. I never at any other time got more than five birds with one barrel, even when properly loaded for ducks.

The *Australian Teal* is a handsome little duck, not quite so large as the teal at home, and, next to the black duck, the commonest of all the species. They generally flew in fair-sized flocks, often mixed with the black duck, were tolerably tame, and we rarely brought home a bag of ducks without a couple or so of teal. It appeared to be more common on the coast than any of the other ducks. The male bird is a splendid mottled chestnut and black, with a very brilliant green neck, while the female resembles the European teal. We saw so few of these handsome birds in proportion to the others, that I always considered it a distinct variety, which some of the old duck-shooters also did, and used to call it the " merganser." But a young friend of mine took the nest, with seven eggs, out of a hole in a gum-tree, and shot both the old birds, a handsome male and a dull female. Still I felt certain we had two varieties, and that all the dull-coloured birds we killed were not females, and in April, 1857, I shot a dull-coloured bird with a red eye, which, on dissection, proved to be a male. Teal fetch about three shillings per couple in

the market, are considered the finest-eating birds of the
whole lot, and a teal supper at ten shillings per head
used to be the general evening's finish for the " men
about town " in Melbourne.

The *Musk Duck*, so called from the strong musky
scent peculiar to the male, especially in the breeding
season, is a singularly ugly bird. Clumsy and chubby
in shape, as large as a small goose, of one uniform dull
grayish-black colour, thick head and beak, and the male
has a large warty flap, or excrescence, hanging down
from the chin. It has a curious appearance when swimming, the body almost entirely under water, the head
and neck alone visible. It was, I believe, not uncommon
in some of the inland lagoons, but rare with us. In fact,
it is a shy solitary bird, frequenting creeks and water-
holes grown up with very thick rushes, and not often
seen. The wings are mere rudiments, like those of the
divers, to which class of birds I fancy it belongs, and it
trusts much more to its powers of diving than flight for
its safety. I never saw one on the wing. I have killed
it out at sea, in the bay, but I generally used to come
upon an odd one in some out-of-the-way creek or water-
hole, and never saw more than two together, although
they bred with us. It is rank and fishy to the taste,
and, except as a curiosity, hardly worth shooting. Some
call it the *Moss* duck.

This completes the list of Victorian ducks, and it will
be found very meagre in varieties, when compared with
that of Britain, which numbers about twenty-six varieties,

exclusive of the swans and geese. There are no true sea-
ducks in this part, but nearly all the species which I
know appear to frequent the salt as well as the fresh
water.

CHAPTER VI.

THE COOT—THE WATER-HEN—DABCHICK—BITTERN—HERON—WHITE
CRANES—EGRET—SPOONBILL—IBIS—NATIVE COMPANION—NAN-
KEEN CRANE—COAST-SHOOTING—SEA BIRDS.

THE Australian *Coot* is the porphyry-bird or sultan-hen
of South Africa, and much resembles the British coot in
size, shape, and habits; but the body-colour is a beautiful
blue and white under the tail; the cere, beak, helmet,
and legs are bright red. One of the peculiarities of this
bird is, that it can bring its food to its mouth with its
feet, which are not lobed like those of coot at home, but
the toes are long and thin, like those of the water-hen.
They were very common in most of the sheltered creeks
and water-holes. They bred with us, and in the autumn
appeared to flock, and then we principally found them in
rushes or tea-tree scrub, in which they perch. They are
very hard to rise, run like lamplighters, are easy to shoot
when on the wing; and though I liked them much when
roasted, were hardly worth shooting for the market.

The *Water-hen* was much rarer with us than the coot.
I generally found them in thick rushes, and never saw
more than two together. It very much resembles its
British namesake in size, habits, and general appearance.

The *Dabchick* here very much resembles the little dab-

chick at home in all respects, but was prettier about the
head. It was a summer migrant to our parts, and a pair
or two might then be seen on any water-hole ; and it is
a wonder how they become so generally dispersed, when
we consider their weak powers of flight. We had one
or two other species of grebe, very rare, however, in our
district.

It is no wonder that a country like this should abound
in swamp birds of every description, and the *Bittern*, which
more than perhaps any other shuns the haunts of man, is
one of the commonest of the wild tenants of the Australian
waste. I have killed eight or ten in the day, rising from
the rushes and grass in one large swamp, and any day
in the autumn I could bring home a couple of bitterns.
They appear much to resemble the European bird, but
are a little duller in colour. The call-note is exactly the
same, and often have I been startled, when quietly seated
at night watching a duck-hole, by the heavy bump of the
bittern from the reeds close to me ; and as the weary
shooter is plodding his homeward way, after evening has
closed in over the dreary swamp, the dull measured boom
of this solitary bird appears to add to the desolation
which reigns over all. I have heard of a little bittern
being killed out here, but never saw one.

The *Heron* is very common on the low marshy grounds,
and by the sides of the creeks ; and I have seen large flocks
of thirty or forty together. In size and plumage it resembles the European heron, but is not nearly so fine or
handsome a bird ; and many of the feathers, especially

the scapulars, have a much redder tinge. It is gregarious in its habits, except in breeding; for, unlike the herons at home, the Australian heron builds a very small solitary nest on some old tall gum-tree, often far away from water. I always fancied we had two species of heron, the one much smaller than the other.

The *Purple-and-white Heron* occasionally came down into our parts, generally in small flocks; but I considered it a rare bird. In appearance it resembled its European namesake; but I fancy it was rather larger and handsomer. It is, I believe, common in Van Diemen's Land; at least, we used to call it the Van Diemen's Land heron.

We never shot any of these birds for the market, but we always ate them ourselves, and, to my fancy, they were fully equal to any of the so-called game-birds.

We had two species of *White Heron*, or, as they were called by the shooters, the *White Crane*,—the one much resembling the great white heron of Britain in size and appearance, with a black beak, and another variety, which was much smaller. I think the large white heron was the commonest with us. Now and then an odd one came on to the large swamps in the winter; but their principal resort was Western-port Bay, and I have seen as many as a dozen feeding together at the mud flats there at low water. It is a shy, wary bird. They breed on the large rocks out at the Heads, and seemed to come down to our district in the autumn. I have, however, seen them in Western-port Bay a very little after Christmas.

The *Little White Egret* was a very rare and casual visitant to our parts. I only saw two specimens killed with us. It seemed exactly to resemble the egret at home, and is, I think, one of the chastest and most elegant birds in the colony.

The *Spoonbill* was rare with us, and I only knew of about three specimens being killed. It is an elegant bird, pure white, with a fine pink tinge under each wing. In some places it is very common.

Occasionally, an odd *Ibis* is killed here; and the specimens that I saw resembled the sacred ibis of Egypt in plumage, and had not the purple tinge peculiar to the ibis of Britain. It is an ugly dull-coloured bird, and has a tuft of curious feathers, like a bunch of coarse hay, hanging from the breast. We used to call it the straw-necked ibis. The real home of the ibis is, however, far inland; and it is only when the up-country is heavily flooded that they visit the districts near the coast.

The *Nankeen Crane*, or night heron, is another chaste-looking bird, and a summer migrant to our parts, coming down in October to breed, and leaving in the autumn. The whole body-colour is pure nankeen, black cheek and head, white belly, yellow eye, cere, and legs, and three long white feathers, so closely joined together as to appear but one hanging down from the back of the head. They were far more common with us in some seasons than others. The nankeen crane is strictly nocturnal in its habits, sitting by day moped up in the high gum-trees or tea-tree scrub, half asleep; as soon as evening sets in,

they wake up to feed, and the hoarse croak of this bird
may be heard about all the swamps and creeks throughout the whole summer night.   They are very easily shot
by day; for, when disturbed, they rise with a heavy
wing, and seem, like the owl, scarcely to know where to
fly, and soon pitch again.

We had another species of bittern, or heron, in shape
and size much resembling the nankeen crane; but it was
of a light chestnut-brown colour, variegated with black,
and had not the long pendent feathers peculiar to that
bird.   It was not so common, seemed to be much more
diurnal in its habits, and I oftener used to see them by
the sides of the creeks than on trees.   I called it the
" spotted bittern," for want of a better name.

The last on our list of the swamp birds, although certainly not the least, is the *Native Companion*, or Australian crane.   This bird is larger than the European crane,
which it resembles in shape and habits; but the colour
is uniform light slate-blue, with a red cere and bare head,
and it wants the handsome tail-feathers peculiar to our
crane.   They are about the most wideawake birds in the
colony; and, as they generally frequent the open swamps
and wet plains in small companies, and the old male bird
is always marching about on the look-out, every now and
then uttering his loud trumpet-like note of alarm when
danger is near, it is next to impossible to stalk them in
the open; but, in the end of summer, they draw down
to the edges of the creeks, and are then easily approached
under cover of the tea-tree.   I once dropped on a little

mob of five in such a place, and I nailed three at a double shot; and well I recollect bringing them home on my back at night, about six miles, with five couple of black ducks and thirteen pigeons. An old bird will stand over five feet high, and weigh upwards of twenty pounds. I once found the nest in a swamp near us: it was built high, of dry rushes, like that of the swan, and in it were two large eggs, mottled with red, especially at the large end. I once caught a half-grown young one, which I kept at my tent a long time. It was a voracious feeder, and lived principally on boiled rice.

There are very few sea birds on these coasts. From Mordialloc down to Frankstone, on Port-Phillip Bay, the beach is low and sandy for eleven miles, and beyond this to the Heads it is high and rocky. The shores of Western-port Bay are principally mud-flats, fringed with mangrove scrub; and on the Williamstown side, as far as Gelong, the coast is low, edged with banks of seaweed, washed in by the tide, and also fringed with the mangrove. On these flats the ducks, waders, and pelicans feed at low water, and two or three species of gull and tern, curlews, avocet, and large flocks of stints, are met with up and down the whole beach. But to the coast-shooter neither of these bays offers much attraction. Further out towards the Heads, the coast is less disturbed, and bluff headlands of ironstone rock afford a wilder and safer home for the sea-fowl; and facing the wide ocean many other and rarer species are probably met with than in our land-locked bays.

CHAPTER VII.

THE PIGEON—THE SNIPE—THE RAIL.

THE bronze-wing pigeon, for size and beauty of plumage, certainly stands No. 1 on the list of Australian bush game; and of this bird we had two varieties,—the common bronze-wing and the scrub pigeon.

The *Bronze-Wing* is a beautiful bird, plumper and larger than the dove-house pigeon, but not so large as the British wood-pigeon; the upper plumage dark brocoli-brown, the breast and neck glossy and shining, the under parts light, the forehead white, and on the wing is a beautiful speculum of bright bronze-coloured feathers, from which the bird derives its name. We had no blue pigeon in Victoria. The male bird is finer and handsomer in plumage than the female, the white on the forehead much larger, tinged with chestnut-red, and I fancy that this tint becomes deeper with age; the cheeks and throat are deep chestnut, the wing-speculum larger and brighter; and a glorious bird does an old cock bronze-wing look, when seated on the bare limb of a large gum-tree, his burnished wings, chestnut head, and glossy breast reflected in the rays of the evening sun.

Like most of the game birds, the bronze-wing pigeon

was a summer migrant to our parts, coming down about
the end of September for the purpose of breeding, and
what few escaped the gun, left about the end of March.
An odd straggler or two would certainly remain in the
forests throughout the winter. At different seasons they
frequent different localities. When they first arrive,
they are to be found among the shey oaks and large
honeysuckles, generally on dry rises, and as often on the
ground under the trees as up in the branches. As the
season advances, they get much into the heather, especially at night and morning; and both the pigeon and the
quail are very partial to heather that has been previously
burnt. They are very fond of the wild cherry. When
the thistle-down is floating, every patch of thistles holds
a pigeon; and as soon as the wattle-trees drop their seed,
you will surely find the pigeon at the foot of them; in
fact, you may look for pigeons in the wattles at all times.
They breed principally among the honeysuckle and shey
oaks; the nest flat, similar to that of the wood-pigeon
at home, in which the female lays two white eggs, and
the old cock-bird takes his turn at sitting. I once found
a nest with eggs as late as February 4th; but I fancy
this was a second clutch: not that I think the pigeon
breeds more than once in the year, but, like the partridge
at home, when the first clutch of eggs is destroyed, the
old female lays a second. By the end of January, the
young birds are strong fliers, and large flocks of pigeons
then congregate in some favourite localities, previous to
leaving; but where they go, or from whence they come to

us, nobody seemed rightly to know. For about a month
from this time, a man who knows just where to look for
them can have some rattling sport. The most I ever killed
in one day was eleven couple and a half; and this was not
an individual day's luck, for pigeons were so thick in the
month of February in that year, in the honeysuckle and
shey-oak scrub on the beach, when I was camped at
Mordialloc, that I averaged with my own gun twenty-five
couple per week for above a month. Although the
pigeons flock here, they generally rise singly; or, if two
or three fly up together, they are so wide apart that you
rarely kill more than one with each barrel, and you never
get a "family shot," as you can into a flock of wood-pigeons
at home. I have occasionally killed two at a shot, young
birds, sitting together on the same branch. The coo of
the pigeon is deep and loud, principally heard at night
and morning, and often leads the shooter up to them in
the forest. The surest but most pot-hunting method of
killing pigeons is to creep up to them as they sit on the
bare limb of a tree; and a dull, warm, rainy day is the
best for this kind of shooting. The blacks are the boys
for this work. A certain way of killing pigeons is to
watch by a water-hole on a summer's night, just as the
sun goes down, when they come to drink; and I have
killed eight or ten in an evening at a favourite hole, and
this in not a very good pigeon country; but they will
come a long way to water. When "roading" woodcocks
in the north, the first appearance of the evening star
was the signal for the shooter to take his stand in the

forest glade; and here, in pigeon-shooting by a water-hole, as soon as ever the evening star shows, you may go home. The most sporting way of killing them is as they rise from the heather, or the ground among honey-suckle scrub, when they go away as straight and as sharp as any of "Barber's best blue rocks." A great country for pigeons is about the Snrney, on the coast, forty miles from Melbourne; and another famous place is on the Gipp's-land road, below Dandenong. But they are to be met with in larger or smaller quantities all over the bush. As the small settlers begin to take up the forest land, the pigeons disappear; for, although I have heard of them in the corn-fields, their principal food is certainly seeds and berries; and, although not a very shy bird the wild bush, the pigeon likes quiet and secluded places to breed in. Pigeons will fetch 2s. 6d. per couple throughout the year, and they are well worth it.

The *Scrub Pigeon* is a smaller, and, I think, a handsomer bird than even a common bronze-wing. It is much rarer, generally found singly or in pairs, very seldom in small flocks, except late in the season; the colour is a uniform dark cinnamon-brown, the forehead reddish, and the wing-speculum, although not so large as in the common bird, is far deeper and more brilliant. It is very partial to particular localities, and, like the woodcock at home, there are certain places where you will always find a scrub pigeon. It is a shy, solitary bird, frequenting the thickest scrub, and seems partial to tea-tree by the side of water. They almost always rise from the ground. We used to kill an odd scrub pigeon at times all through the winter; but about April and May, when they congregate, is the best time for shooting them. In fact the best season for them appears to be after the other pigeons have left.

We had a little bird on the ranges which we called the *Ground Dove*, about the size of a fieldfare at home, and much more like a thrush than a pigeon. It was a summer migrant to our parts, came and left with the painted quail, and was generally to be found on the ground on dry rises in the forest among fallen timber. It rises with a loud flutter, and flies with a dipping kind of flight. It is a pretty bird, variegated red, brown, and black, with chestnut markings, and five or six white diamond spots on each wing-shoulder. It lays on the ground three largish, mottled, reddish eggs, in a careless nest. Although not strictly game, we used to sell them with the quail.

There is a large species of pigeon on the Sydney side, called the crown pigeon, but it is not met with here.

The Australian *Snipe* is much larger than the common English snipe, shorter in the leg, plumper, and thicker; and the general plumage and appearance, its manner of rising and flight, remind us more of the double or solitary snipe of Europe, than our common bird. There is no real woodcock in Australia, and I fancied that the snipe here appeared in some slight respects to partake of the nature and habits of that bird. I never saw a jack-snipe out here, nor do I believe there is one, although some

shooters say that they have killed them; but I think
this was nothing more than a kind of little stint, which is
often found on the plains. Where the snipes spend the
winter and breed, no one seems to know. I have heard
that they breed on the high ranges at the head of the
Yarra, and a friend of mine has flushed them in June
in the Stringy-bark ranges, 200 miles up the country.
One thing is certain, they must breed very early; for
when they came down to us in September, there was no
difference in the size of the birds that we killed; and I
believe there can be no doubt that they did not visit our
parts till after the breeding season, for I never heard
of the nest being taken, and the habits of the snipe that
came to us were not those of breeding birds. They
appear in the districts round the coast in September,
remain throughout the summer, and leave in February
or the beginning of March. They come down by stages,
for we generally heard of the first snipe being killed
up country a fortnight at least before they reached us.
The first place that they visit in our district was the
Clyde, a low flat of wet pasture-ground, about fifteen
miles below Dandenong, towards Western-port Bay.
This is the best and earliest snipe-ground that I know;
but the water very soon goes off, and a man, to have any
good shooting, should be there when they first come.
They then take another flight, and, like the snipe at
home, following the flood, come into the Dardenong
country, and thence disperse themselves over the
swamps and low grounds, frequenting of course peculiar

H

localities where there is good feeding-ground, till they
reach the coast, where all that are spared remain
until they leave; and I could always make sure of
a couple or two in the honeysuckle or tea-tree scrub
along the beach, when I could find them nowhere else.
The habits of the Australian snipe are very puzzling,
and a man who is not used to snipe-shooting here may
beat acre after acre of what we should consider in
the fen capital snipe-ground, without springing a bird,
and perhaps pass over the very places where the snipe
do lie.   Fancy an old fenman trying for a snipe among
ferns and heather on a dry sandy rise, or in thick honeysuckle scrub; yet these are the very places to look for
the Australian snipe: in the summer and in the heat of
the day you will find them here in large wisps, and nowhere else.   In the early part of the season a man may,
however, beat for them in much the same places as he
would at home; and as the season advances, they lie much
under the shelter of any large timber near the swamps,
and in patches of tea-tree which skirt the creeks and
wet ground.   They never lie far in, and an old dog who
knows his business will potter steadily along a yard or
so in the tea-tree, and tumble out the snipe as fast as
ever you can load and fire.   In the very heat of summer
they get much into the honeysuckle scrub, but always
somewhere near their feeding-grounds; and here it is
snap-shooting with a vengeance; for when they rise they
are only seen for an instant.   The Australian snipe in
the open is not nearly so difficult to kill as the snipe at

home. They are a larger object, fly much steadier, and generally go away straight; yet, owing to the places they frequent, are often missed. They are very fond of lying in the shade by day. If by chance any large gum-trees stand in an open wet plain, they will generally get under them, and I have often planted myself under a favourite tree, and stood still while others were beating the ground round me, and killed as many as all the other guns. They usually rise quietly, but I have heard them "scape" like the English bird, especially when coming down to the feeding-grounds at night. I fancy one wisp follows another as they are travelling down, for in some days you will find snipe in places where a week before there was not one. Of course, this is much owing to the state of the feeding-grounds and the season; before the water dries up, they are dispersed over the whole face of the country; but as it goes down, and many of the feeding-grounds become parched up, they pack more. There are then certain places where you are always sure to find some, and a man must know the country well who can make sure of a bag of snipe late in the season; for I never knew a bird that sticks to favourite localities more than the Australian snipe. They shift their quarters in the early part of the season very suddenly, and if a man hears of a wisp of snipe in any particular place, he must be off at once, or, upon reaching the ground, he will probably have the mortification of seeing the feeding-marks of hundreds of snipe, and find perhaps only a few outlying birds. The Australian snipe is a terrible bird

to run, and you will rarely rise one just at the spot where
you saw it pitch. They often perch in the tea-tree scrub,
and I have twice killed them sitting on the bare limb of
a large gum-tree. Whether for sport or profit, I consider
the snipe the finest small-game bird in Victoria. They
remained in our district longer than any other summer
game. There is no pot-hunting in snipe-shooting, they
must be killed in a sportsmanlike manner, or not at all.
It is fair to shoot them whenever they are found. Every
one knows the pleasure he experiences in a good day's
snipe-shooting, and what was of the most consequence
to us, we had always a ready sale for them in Melbourne,
at 2s. 6d. per couple; and occasionally some free-liver
will give 5s. in the first of the season: in 1853, I sold
the first snipe that I killed for 5s. Although this is
certainly a great country for snipe, yet I have never
seen such wisps here as in Sweden, when the old and
young birds were on their way down from their northern breeding-haunts in September. The most I ever
bagged here myself in the day was thirteen couple and
a half; and although I have heard of some extraordinary days' snipe-shooting, I never myself saw twenty
couple of Australian snipe fall to one gun in the day.
No bird has been driven from this district more than the
snipe, and to get a good day's shooting a man must now
go a long way afield.

As a specimen of a day's sport out here, I will give an
extract from my game-book of December 22nd, 1854, on
which day "the old boy" and myself shot on the island

near Mordialloc. We both shot well and pretty even, and all was game on that day. At night we brought home to my tent—

16½ couple quail, 3½ couple scrub quail, 1 rail, 3 couple pigeons, 11 couple snipe, 3 nankeen cranes, 1 red lowry, 5 black-ducks, 3 shovellers, 3 coots, 2 black cockatoos, 2 moorhens, 7 shell parroquets.

I do not quote this as anything extraordinary, and I have no doubt it has often been beaten; but I fancy it would puzzle two men to do it again on the same ground. It will, however, give the reader an idea of the varied contents of an Australian game-bag.

The painted snipe is the common snipe on the Adelaide side, but is not met with here. It is a pretty variety, and something resembles the painted quail in plumage.

The Australian *Landrail* is a species of crake, as large as the corn-crake at home, but handsomer in plumage, and principally frequents rushes and sedge in moist situations; but you often find them in fern on dry rises, a long way from water. They are very common during the summer, very hard to rise, run a great deal, fly exactly like the corn-crake at home, and their cry when disturbed is a sharp " chip, chip, chip." They are excellent eating. Bred with us, and left early in the autumn.

We had two smaller varieties, which I have described hereafter, in my notes on the ornithology of this country, as they scarcely come within the list of game birds.

I can't say that ever I shot a true *Water-rail* out here. I have killed a small dark-coloured bird in the swamps rather resembling that bird, but I do not remember whether it was a rail or a crake. One thing is certain, if there is any real water-rail in this country, it must be very rare.

CHAPTER VIII.

QUAIL-SHOOTING—THE COMMON QUAIL—THE SCRUB QUAIL—THE PAINTED QUAIL—THE NUTHATCH QUAIL—THE KING QUAIL—THE SPUR-WING AND OTHER PLOVERS—A HINT TO BIRDCATCHERS.

THE *Quail* is the Australian partridge, and quail-shooting is certainly the least laborious and pleasantest of all field sports out here. It reminds the sportsman of September at home, for it is fair open sport, and a man can have the pleasure of seeing his dogs work in the old style. Moreover, they are generally pretty thickly dispersed over the whole country, and in a few hours' shooting a tolerable shot can always make a nice little bag.

We had six varieties of quail in our district,—three common and three rare : the common quail, the scrub or partridge quail, the painted quail, all common in their peculiar localities ; and the nuthatch, the king, and the silver quail, all rare and only occasional visitors.

The common quail comes down about the middle of September, remains to breed, and early in February they all appeared to leave the breeding-grounds, but not the district, for they then packed, and in certain localities large flocks might be seen late in March ; but after March we rarely saw a common quail in our parts. I have observed that the quail leave the heather sooner

than the grass. Where they winter nobody seemed to
know, but I fancy they go back into the large plains in
the interior, from whence they appeared to come to us;
for if they had come over the sea, we should have always
found them on the coast first, and many would have been
picked up on the beach in a state of exhaustion, like
woodcocks at home. I observed when they first came, a
few birds would arrive, as the pioneers, perhaps a week
before the great flock; and one thing which surprised
me was, that you might beat the same ground day after
day, and, however many you shot, the number of the
birds did not appear to diminish. When they first arrive,
they are generally to be found in the long grass on the
edges of the swamps, on the grassy plains, and in heather;
and these are the general places to beat for the common
quail throughout the season; but as the corn springs up,
they draw much into the cultivation paddocks, where
they breed in security; and the quail is the only game
bird here that is likely to increase with population. In
the hot summer they are always to be found on moist
ground and in the neighbourhood of water-holes, especially at mid-day. They feed at morning and night, and
the best time of the day for quail-shooting is from three
in the afternoon till sunset. In the early morning, when
the dew is on the grass, they won't lie, and in the middle
of the day they lie too close; and as there is then no
scent, the dogs are almost sure to pass over them; moreover, the dogs can't hunt here in the heat of the day.
It is next to impossible to rise quail without a dog;

three men in a line, beating the ground slowly, may get up some, but they will walk over far more than they spring. Quails squat very close and run very quick. A close-hunting, heavy retrieving spaniel would be the best dog in quail-shooting here, for they require a good deal of bustling to get them up, and this is not a country for a fine-broken pointer; for, owing to the running of all the game birds, and the quantities of field-mice that infest the plains and heather, I'll defy any dog, no matter how well broken, to be staunch to his game out here.

The great drawback to shooting small game in this country is the quantity that is spoiled by the heat. A large fishing-creel is the best thing to carry small game in, packing them carefully in on layers of grass or tea-tree, as we serve the grouse on the moors at home. As soon as you come home, wipe away all the blood and loose feathers, and hang the birds in small wisps up in a draught: the higher they are the more they will be out of the way of the flies. An old friend of mine used to adopt a capital plan with his snipe and quail. As soon as he came home, he tied up each bird separately in a cabbage-leaf, and laid them carefully in an iron camp-oven, keeping on the top. No English sportsman can form any idea how soon the game goes here. The flies blow so quickly, that I have often taken a bird out of my bag, killed but a few hours, a living mass of maggots. It is a good plan, if your day's sport keeps you in one spot, to hang the birds in small wisps as you kill them, high up in tea-tree and other scrub, in the shade: they soon spoil if mashed about in a pocket or game-bag. A little pepper in their mouths and vents freshens them. It is not the man that shoots the most game out here who makes the most by it, but he that takes the best care of it.

The *Common Quail* is a pretty gamy little bird, very much like the European quail in size, habits, and appearance; but I fancied it was prettier. The call-note, when on the ground, much resembles the native name of the bird, "too-weep," often and loudly repeated, especially when feeding: the cry when they rise is a sharp chirp. Although a small object, the quail is not a difficult bird to kill, on account of its straight flight.

We used to kill a large variety of the common quail, which we called the *Stubble Quail.* It was rarer than the common bird, larger and thicker; the breast of the male, instead of being black, was plain-coloured, and there was also a slight difference in the beaks.

Quail-shooting is not a bad game where a man has regular customers. I used to consider from fifteen to twenty couple a good day's work (I once killed thirty-seven couple), and I rarely bagged more than fifteen out of twenty, taking in misses and lost birds. A man soon empties his flask in quail-shooting, and ammunition is no slight item in the expenditure of the small-game shooter out here: I reckoned every couple of birds cost me 3*d.* to kill, and they averaged 1*s.* per couple throughout the year. Although always found in "bevies," quail gene-

rally rise singly, or quickly one after the other, and never,
like partridges at home, in coveys.

The best season's quail-shooting I ever knew was when
my old mate Rendall, or " the old boy," as we called him,
shot on the heather at Picnic Point, about twelve miles
south of Melbourne. He bagged 1,500 couple of quail
on one ground in the season; but he had miles to shoot
over. Twenty-five couple per day was his general bag;
he averaged eighteen birds out of twenty shots, and he
used to work at it day after day, like any other kind of
labour. But he certainly was the best shot I ever saw
take a gun in hand (and I have shot by the side of " the
Squire " and other good men), and there was scarcely
his equal in the colony in beating for game. He shot
to a couple of little mongrels, the smallest a bobtailed
terrier, about 5 lbs. weight, and " Johnny " rarely passed
over a quail. I never used setters or pointers in quail-
shooting; our dogs were up to every kind of bush-work,
from driving a kangaroo to hunting for quail. Of course
there are plenty of well-bred setters and pointers out
here, and we generally see the best dogs in the hands of
men who use them least; but the Melbourne sportsmen
can now, as the advertisement runs, have " their dogs
broke as they ought to be, by a Leicestershire sportsman,"
at £5. 5s. per head.

The common quail is found on one of the New Zealand
islands, but I believe there is no snipe in that country.

The *Scrub Quail*, or, as we called it in the bush, the
partridge quail, is the largest of all the species, with a

fine brown mottled and barred plumage, like the gray-
hen at home. We had two varieties, the one much larger
and darker in colour than the other. The scrub quail
rises like the partridge, flies strong and quick, and is decidedly the most sporting bird of the lot. It is nowhere
very common, always in cover or small scrub, in pairs or
families, and in hot weather they lie much on the edges
of the tea-tree by the creeks; and here it is quick work
shooting them, for they invariably rise towards the scrub,
and are out of sight in an instant: three or four couple
of scrub quail in the day was good work in these parts.
Unlike the common quail, they appeared to remain with
us throughout the winter. The common quail lays from
six to eight largish eggs on the ground, very deeply
blotched with reddish brown at the large end: both the
scrub and painted quail lay fewer, the eggs of the former
being white, those of the painted quail light speckled.

The *Painted Quail*, or *Wanderer*, is the handsomest
of the three, and, as its name imports, the plumage is
prettily variegated or painted with red, white, and black;
the legs are yellow, and it has but three toes. It is
intermediate in size between the two last, and the flesh
is whiter. Although you may occasionally kill an odd
one during the winter, the majority of them come in
September, and leave in March. The painted quail is
rarely found in the open, but generally in timber on
ferny or heathery rises. They run very much, have a peculiar wavering flight; and I consider the painted quail,
in timber, as difficult a bird to kill as any in the colony.

They do not pack, like the common bird, but, like the scrub quail, are always found in pairs or families. The note of the male bird much resembles the cooing of a pigeon, but is not so loud, and always repeated twice quickly; and this monotonous call may be heard in the forest throughout the whole summer's night. It is more common than the scrub quail, and when the young birds are fliers, a man has no trouble to kill five or six couple; for when flushed, they soon drop again. The wings are long and pointed, unlike the full round wing of the two last species.

The *Little Nuthatch Quail* was a rare and uncertain visitant to our district, but is, I believe, the common quail on the Adelaide side. I always found them in the heather with us, singly or in pairs, and I scarcely ever killed more than a couple in the day. Like all other partial migrants, they were much commoner with us in some years than others; but it certainly was a rare bird in our district. It is not so large as the common quail, of a uniform yellowish stone-colour, mottled with black and white; the beak large, and unlike any of the others in shape; the legs yellow, and the toes three in number; and, from the pointed wing, I consider it closely allied to the painted quail. What few came into our parts appeared to breed with us; and if so, they left the earliest of any.

The *Silver Quail* was very rare with us, and I only saw two examples, both skins, and both killed on the plains near Melbourne. It appears to be much like the painted

quail in size and form,—a long loose-feathered bird, with
pointed wings; but it is much lighter in colour, and has
a kind of dark collar round the neck. Respecting this
bird, all I can say is, if it is a distinct species (which I
doubt), it must be very rare; for, during five years'
shooting, I never met with a single specimen.

Last and least on our list is the little *Chinese* or *King
Quail*, which, although small in size, for beauty of plumage
stands unrivalled among the game birds of Australia.
Scarcely so large as the common sparrow, a perfect
partridge in miniature, I think we may reckon it as the
smallest game bird in the world. The male is of a deep
velvet-black colour, with rich red chestnut and white
markings, and a dark crescent on the breast; the female
and young birds are deep brown mottled, like the European grouse. It was not common in our districts, and I
generally found them in pairs or families (for they bred
with us, and, if they did not remain all the winter, they
left for a very short time), in the long grass on the edges
of the swamps, often in the wet swamps themselves, and
I have occasionally raised them in the heather. In some
seasons they appeared to be more common than in others.
It is a very local bird; and one thing always puzzled me
in beating for game out here: there are certain localities
where you are almost certain to find birds; while in other
places, precisely similar to all appearance, and apparently
just as well adapted to their habits, you never see a
bird. All the game in Australia appears to pack very
much.

We had two species of plover common with us throughout the year,—the *Spur-wing Plover*, which is analogous
to the *Jacana* of South America, on the low swampy
grounds, and the *Plover of the Plains*, on the open stony
plains and high dry rises. The spur-wing is a fine bold-
looking bird, considerably larger than the British lapwing,
congregates in flocks, and is always to be found on wet
ground. It is a curious and handsome bird in appearance; the body quaker-brown, the breast white, the head
and points deep black. It has a large bright-yellow cere
or flap over the eye (which is also bright-yellow), cheeks,
and forehead, and a large sharp spur, like a cock's spur,
on the elbow-joint of each wing, which I fancy must be
used by the birds for some other purpose than that of
mere defence. The spur lengthens with the age of the
bird: I have seen them, in an old male, nearly an inch
long. They are a very shy, wary bird, difficult to get up
to, have a loud shrill call; and many a shot at ducks have
I lost when, creeping up to a mole on the swamp, I have
chanced to disturb a spur-wing plover.

The *Plover of the Plains* is about one-third less than
the spur-wing, congregates in large flocks, and is, I
think, altogether a commoner bird in its peculiar localities. It is something like the spur-wing in general
appearance, but the colours are not so well marked; the
colour of the body being shiny brown, the belly white,
and it has no spur on the wing. Moreover, it has no
flap over the cheek, but merely a red wart, or lobe over
each eye. The plover of the plains frequents the most

desolate open stony rises and plains so common to this country; is a noisy, restless bird, in habits much resembling the British lapwing; and as they fly round the shooter, they wake the echoes with unvarying cries; and their wild desultory call-note is peculiarly adapted to the barren regions which they frequent. Neither of these birds are strictly game, but we could often sell them at 1s. and 1s. 6d. per couple.

I have not the least doubt that the English partridge would thrive well in the cultivated districts here; in fact, I should think this was the very country for them, and on account of the vast quantities of ants, they could at all times obtain a good supply of food. I do not think the quail eats the ant's eggs. The pheasant has been imported from its native home; but, I believe, has as yet only been confined to aviaries. I do not consider this country nearly so well adapted to the habits of this bird, or any of the grouse tribe, as to those of the partridge. The absence of the pine and larch in these forests would be much against the habits of the pheasant in a wild state, and I do not know what seeds or berries in these forests would supply them with food; for we have no acorn or beech mast here. That they can obtain food in a wild state is, however, proved by the fact of a cock-pheasant being shot within a few miles of Melbourne, out of a patch of tea-tree, a few years ago. It has been turned out loose in New Zealand, and, in one estate I believe, they are fast increasing. As to the grouse, although there are miles of barren moorlands in most parts of this country, the Australian heath does not appear to be at all the same as the bonny brown heather of Scotland. There is a kind of disease peculiar to the poultry out here, which sometimes sweeps off thousands; and I recollect one summer finding great quantities of the little green paroquets lying dead in the forests, which had died from some epidemic.

The *Golden Plover* here is precisely the golden plover of Europe, but much smaller. It was rare in our district, and I never saw them in flocks, but generally in small wisps of five or six. They did not breed with us, but came only at uncertain periods.

The large *Norfolk Plover*, or *Stone Curlew*, was not at all rare with us at certain seasons, in small flocks, but they did not breed with us. They frequented the small belts of timber on the edges of the plains, and I never saw them in the open. They appeared exactly to resemble the British bird. They seemed to be very nocturnal in their habits, and the long melancholy whistle of the stone curlew in the Australian forest at night, often strikes a chill in the heart of the benighted traveller; for an imitation of the call of this bird is a signal-whistle from the bushranger here to his mates at night.

I know no country where a good birdcatcher could do better than in this, and if I had a friend in the line, I would advise him to pack up his traps and be off to Melbourne at once. Quail, plover, and snipe might always be caught for the market, during the season, by any one who understood the business. All the ground paroquets

I

and others could be easily taken in clap-nets, and would have a ready sale for match-shooting. At the pigeon-matches here, five shillings per couple is the usual price for pigeons, and many more matches would be shot but for the dearness and scarcity of the birds. On the Adelaide side the little shell or zebra paroquet is bought up at sixpence each, and much used for trap-shooting. All the handsome parrots, and every species of pretty small bird could be sold in town for cage-birds. I scarcely ever went up to Melbourne from the bush without being asked for live birds or animals; and if I had only understood the trade as well as one of our " Whitechapel bird-catchers," I would have cut the gun and stuck to the net, and nothing else.

The wattle-bird, although not strictly game, will often fetch five shillings per dozen in Melbourne. They come in thickest just as the quail have left, and a man may shoot two or three dozen in the day with ease, for they fly in large flocks, like the fieldfare at home, about the large honeysuckle and gum trees. Parrots can also at times be sold, when game is scarce; and let me say that a parrot pie is no bad dish.

CHAPTER IX.

A CHAPTER ON THE ORNITHOLOGY OF PORT PHILLIP.

HAVING described those birds which more particularly belong to the sportsman, a slight glance at the other species most commonly met with in the Melbourne district will perhaps not be without interest to the general reader. But I may as well at once state that I have neither the intention nor ability of entering upon the subject scientifically. The few remarks that I am about to make are solely the result of my own observation, for I had little or no assistance in my zoological researches out here. I had no work on the ornithology of the country to guide me, and no one who knew the birds to help me. I know nothing of the Latin names of the birds, nor to what class even many of them belong; and the English names which I use are those by which they were known to us in the bush, and perhaps many of them altogether wrong. My notices must necessarily be short and very imperfect; and, as I had not the slightest intention of publishing when in the bush, I kept but few notes, and nearly all that I have written is from memory. I have, however, as far as I could, endeavoured to give a description of such birds as I know; and, short as they are, I trust my notes will answer the purpose for which they are intended.　To enable the stranger to form some slight idea of the ornithology of this country, and the bushman, if he cares at all about it, to distinguish one bird from another, I have noticed above 180 different species which have passed through my hands, and, with the exception of less than a dozen, I have shot specimens of every one myself.

No one has better opportunities of studying nature than the sportsman, whose life is spent in out-door pursuits; and if such men would only pay a little attention to the subject, and note down anything that struck them as worthy of notice in the habits of the animals and birds which are constantly before their eyes, what a fund of useful information might be collected.　But, unfortunately, it rarely happens that either the sportsman or gamekeeper cares anything except about those very birds or animals which are the immediate objects of their pursuit, and scarcely even know the names, much more the habits, of the commoner species, which are of no value for the chase.

The study of ornithology has always been a favourite one with me, and is perhaps the only one of the innocent pleasures of youth which follows a man into maturer years, and upon which he can look back, in the decline of life, with feelings of pure and unalloyed joy.　The greatest charm attendant upon this study is, that there is no monotony in its pursuit,—no void or blank in the ornithologist's year.　His time is constantly occupied; as soon as one class of birds leaves, another arrives; and these migrations are, without

doubt, the most wonderful of the many wonderful phe-
nomena in nature. Instinct here stands forth clear and
unguided, and the actions of the birds themselves arise
from causes over which they can have no control. So
beautifully and with such precision are they arranged,
that we can time the arrival and departure of our regular
summer and winter migrants almost to a day ; and each
particular class is the harbinger of a particular season.
All this is far more apparent in northern countries,
where the vicissitudes of climate are more sensibly felt
than in the warmer latitudes of the south. Let us turn
for awhile to England, and here we shall find that the
opening of the first violet in the sheltered bank of the
village lane welcomes the first spring migrant to our
shores ; and no sooner do the rude blasts of autumn
sweep through the forest glade, whirling the dead leaves
on high, and shaking the last tottering acorn from the
oak, than the chattering of the fieldfares high in air, and
the keeper's report that he has flushed the first woodcock
in some favourite spenny, warn us that winter is again
at hand. The very operations of the husbandman and
sportsman are in a great measure regulated by these
migrations. They form a useful and instructive guide to
the farmer, who will take the trouble to observe them,
and the appearance of the swallows on some favourite
stream, whither in early spring they dash backwards and
forwards over its margin after the "glad May-fly," just
awakening to its ephemeral life, or when, in the haze of
an autumn evening, they congregate in flocks on the

osiers that fringe its banks, is hailed with equal delight both by the contemplative angler and more boisterous huntsman; for each hails it as a joyful omen that his season has again come round.

All this is much more marked at home than in a foreign land, where the birds are strangers to us, with whose habits we have hardly had time to become acquainted; but the same remarks will apply with equal accuracy both to England and Australia. It is true that in this latter country, these migrations being more partial, are far less observed, and are perhaps instigated in some respects by different causes; but the two principal causes are doubtless the same here as elsewhere: search after food, and suitable localities for the purposes of breeding. The advent and departure of the quail, the pigeon, the snipe, and the other regular summer migrants, are conducted with the greatest regularity, and the partial migrations of the large flocks of parrots, wattle-birds, and others, which are constantly taking place, are no doubt regulated by the state of the blossoms and seeds upon which they feed. The more attention that we pay to this subject, the more regular shall we find these migrations, and many a useful lesson, both in the botany as well as the rural economy of this land, might be learned by observing the habits and noting the migrations of the birds to and from each particular district.

Man's constant companions in every out-door occupation, cheering him with their plumage or their songs, affording him often a principal means of subsistence, it

is little wonder that the study of the habits and natural
instincts of birds should be a favourite one with all; and
to that man whose time is happily and quietly spent in
the forests and the fields, it gives one of the truest zests
to rural life.

Victoria is very rich both in species and individuals
of the hawk; and this is not to be wondered at, when
we consider the wild nature of the country, abounding
as it does in every kind of food peculiar to the birds of
prey.

The king of birds here is certainly the *Eagle-hawk*, or
Wedge-tailed Eagle, which, although inferior in size and
attributes to the golden eagle of Europe, is nevertheless
a fine powerful bird, and the largest bird of prey in the
colony. The eagle-hawk varies much in size and colour.
Whether this is owing to a difference in age or sex I am
unable to say, but I fancied we had two distinct species;
the one very dark brown, nearly black at a distance, the
other much lighter in plumage (I have seen one as light
as the European kite); and the two birds, in difference
of colour, resembled the golden and white-tailed eagles
of Europe, but the eyes of both were dark. The dark
variety of our eagle-hawk was the rarest with us, and
was a thicker and shorter bird than the other: the tail
of this bird is long, and in the form of a wedge, which is
very apparent when it is in the air. They were by no
means uncommon in our district at all seasons, often in
pairs, both in the deep forests and on the plains, over
which they would soar almost out of sight, round and

round in steady circles, without apparently moving their
wings. We had plenty of them on the kangaroo-ground,
and I procured above a dozen fine specimens in one
winter. They were often on the ground, and I fancy
were principally carrion-feeders; they bred in our
forests; the nest very large, invariably placed in the
fork of a large gum-tree; not always very high, but
generally inaccessible to any but a black. Several old
deserted nests stood in the forests, mementoes of bygone days, before the foot of the white man trod these
wilds; and I recollect the eagle-hawk's nest on an old
blasted gum was one of our favourite " trysting places "
when driving kangaroo; this bird is not nearly so shy as
the European eagle, and when gorged with carrion by no
means difficult to approach.

The *Large White Fishing Hawk* was by no means
rare on our coasts; they were generally flying up and
down the beach, and I rarely saw them far inland. It is
hardly so large as the wedge-tailed eagle, but thicker and
more robust in appearance, and rounder in the wing
when flying; the tail is not so long, but also wedge-
shaped and rounded; the body-colour, and wings, are
slate-blue; the neck, breast, and belly, white; the shaft
of each feather dark. It was not a true osprey, but in
the shape of the head resembled that bird; the feathers
on the neck were shorter: it was by no means so common
as the eagle-hawk. I once found the nest of this bird on
an old dead gum-tree, in a wood about half a mile from
the coast. We went several times by day to shoot the

old birds without success; at length, one moonlight night I found the tree, and sat under it till morning: just before daybreak the old bird came to the nest, and I shot it. This is the plan I would always adopt if I wanted to shoot an eagle at nest; for it is almost impossible to approach the nest by day without being seen by the old birds. The cry of this bird is a loud hoarse scream.

The *Peregrine Falcon* was common on our plains in autumn, but I do not fancy they bred in our district; it exactly resembled the British peregrine in size, habits, and appearance, and seemed to be precisely the same bird: the eye was dark.

The *Hobby* was also common in the autumn; I generally found it in thicker timber than the other hawks, and I think its principal prey was pigeons; it very much resembled its British namesake in appearance, but seemed to be a little larger: the eye was light hazel.

We had a smaller variety, which we called the *Merlin*, but it was not much like the merlin of Europe; it was common on the plains and in the low scrub during the small-game season.

The Australian *Sparrow-hawk* is about as large as its European namesake, which it much resembles: it was common with us throughout the autumn: the eye was light yellow.

We had an elegant little falcon, not unlike the sparrow-hawk in appearance, but nearly double the size, and much prettier; we called it the *Blue Falcon*: the head and

upper plumage light blue; the under parts barred and
striped with a reddish tinge; the eye bright yellow. It
was not very common; was swifter on the wing than any
of the other hawks; and I generally used to find them in
the end of summer, dashing down the creeks, I suppose
after the ducks.

The Australian *Kestrel* something resembled the female
kestrel at home, and the sexes did not differ in plumage;
it was, however, rather smaller; it was common with us
during the quail season, and generally to be seen in
pairs, beating or hovering over the plains, after the
manner of the British bird, or perched on a dead tree,
apparently watching the shooter.

We had a very pretty variety of kestrel, which we
called the *Little White Hawk*, rather larger and thicker
than the common kestrel, which it much resembled in
habits; it was, however, more common with us, and I
used always to find them beating over the swamps and
low marshy ground, and I fancy their principal food was
reptiles and snipe. The wings and back of this bird were
deep slate-blue; the under parts pure white; the eye
red; cere and legs yellow. It was an elegant-looking
bird, and we generally saw two or three together. They
bred in our neighbourhood, the young birds of the year
prettily mottled, chestnut, red, and white. They left us
late in the autumn for a short period.

The *White Goshawk* is by far the chastest in appearance of all the Australian hawks; about the size of the
European goshawk, but more slender in shape; the

whole plumage pure white, with a bright yellow eye, cere, and legs. I only killed one specimen in our district, and this was by a water-hole; but I have heard they are common in many of the gullies where the native pheasant abounds.

We used now and then to kill a beautiful little hawk,— the *Musquito Hawk*, a perfect sparrow-hawk in miniature, but little more than half its size. It was the smallest hawk I ever saw. It was by no means common, and, like all the smaller hawks, appeared to come into our district with the small game, and leave in the winter. I fancy the hawks here must breed early. Very few bred with us, and many which we killed in October were young birds of the year.

I twice saw a splendid hawk beating the heather for quail late in the season, but I could not shoot it. It seemed a species of harrier, as large as the common buzzard, and was of a rich variegated colour, chestnut-brown, black, and white.

One of the commonest of all the hawks with us was the large *Marsh Harrier*, or, as we used to call it, the *Swamp Hawk*. Throughout the whole time that the ducks were on the swamps, this bird was beating over the grass and reeds; and we often saw as many as half a dozen together flying over the same swamp. I used to kill two species of large hawk on the swamps, the one resembling the British rough-legged buzzard, the other the marsh harrier: this latter bird was much lighter in plumage, and altogether a larger, thicker bird than the

other, which was very dark-coloured, and in the head
and face resembled a harrier. The lighter bird was the
commonest with us. Both used to beat the swamps in
company, and we always shot them when we had a
chance, on account of their killing so many ducks; and
we called both the swamp hawk. I am not, however,
certain that they did not do us often as much good as
harm, on account of driving the ducks up out of the
thick tea-tree and other places in the swamps which we
could not get at. The eye of the darker bird, unlike
that of the marsh harrier, was deep brown.

The bird which we called the *Australian Kite* rather
resembled the British kite in shape and colour, but the
tail was quite square and the rump white. It principally frequented the swamps and low ground; but we
sometimes found it in timber, where I never saw the
swamp hawk. It did not soar so high as the kite at
home; nor do I believe that it was a true kite, although
we called it so. None of the hawks in this country appear to soar very high, except the eagle-hawk.

The *Carrion Hawk*, as we called it, was perhaps the
commonest of all, about half the size of the marsh harrier, of a dull brown colour, relieved with yellow, and a
dark eye. I generally found them in small timber all
over the bush, often in the small belts near the plains.
It appeared to be the most sluggish of all the species,
always gorged with carrion, and altogether the ugliest
hawk I know.

I once shot a hawk as it rose from the heather, when

I was beating for quail, very much to my eye resembling Montagu's harrier. I never killed more than one of this species with us; but I believe it is a common hawk on the Sydney side. It was a true harrier.

Most of the hawks came into our district when the quail set in, and left late in the season; but we saw comparatively few in the winter.

We had at least six species of owl more or less common in our forests.

The largest, which was very rare (the only two examples I ever saw were both killed in the tea-tree scrub by the Dandenong Creek, on two separate winters), was nearly the size of the European eagle-owl, but without horns. It was of a light gray colour, mottled black, with a hawk-like beak and very sharp claws. I know nothing of its habits, except that I have occasionally seen a large owl (which I took to be this) flit by me at night when flight-shooting. I do not think the large owls in this country have any peculiar hoot or cry at night; certainly nothing like the eagle-owl or wood-owls of Europe.

I have killed two species of *White Owl* here, both out of honeysuckle-trees on the plains in the quail season. Neither were common, and they appeared to be irregular summer migrants to our parts. The largest variety was pure white in colour, irregularly ticked and spotted with brownish black; the other was smaller, had a very yellow tinge, and much resembled the barn-owl of England. Both had dark eyes. I never saw either in the winter.

The *Large Grey Owl* was by no means rare, and seemed to remain in our forests throughout the year. It was larger than the wood-owl at home, of a light gray ash-colour, with bright-yellow eye. I generally found it in thick tea-tree scrub in the gullies. The two brown owls were both much smaller, neither of them so large as the common short-eared owl at home; and one was considerably larger than the other. They were both deep cinnamon-brown, the smallest rather the darkest in colour. Neither were rare in our forests, and both remained with us throughout the year.

I never killed an eared owl out here, and the other owls were not nearly so common as I should have imagined, considering the wild wooded nature of the country.

As soon as the shades of evening close in over the Australian forest, the ear is startled by the cry of "morepoke," clearly and loudly repeated, and a bird as large as an owl flits by on noiseless wing, like the goat-sucker at home. This is the *Morepoke*, a species of large night-jar, all head and mouth, about the size of an owl. It is a singular-looking but rather handsome bird, of a deep slate-gray colour, ticked all over with black; the feathers long and pointed, an eye of the most brilliant yellow, and a long pointed tail. The beak and feet resemble those of the European night-jar on a large scale, and the gape is tremendous. It was by no means uncommon in all the deep forests, generally single, and rarely seen by day. They bred with us in the hole of a tree. Their principal food appeared to be large night-moths, and in habits they very much resembled the night-jar at home.

We had a smaller species, which we called the *Little Morepoke*, a rare and pretty little bird; the body not much larger than a lark; the plumage light gray, ticked and barred with black; the feathers soft, the head much rounder than that of the large morepoke, and the tail long and square. I don't believe it was so very rare with us; but on account of its size and habits, not often seen; and it appeared to be very little known among the naturalists here. It was nocturnal in its habits, although the few specimens I killed were by day, as they flew out of a hole in a tree or log. They bred with us, and both species were met with in our forests throughout the year.

We had also a real *Night-jar*, precisely similar to the home bird, which I always used to kill as it rose from the heather, or thick low scrub, in the quail season: it was by no means common, and appeared to be a summer migrant.

About an hour before sunrise the bushman is awakened by the most discordant sounds, as if a troop of fiends were shouting, hooping, and laughing round him in one wild chorus; this is the morning song of the *Laughing Jackass*, warning his feathered mates that daybreak is at hand. At noon the same wild laugh is heard, and as the sun sinks into the west, it again rings through the forest. I shall never forget the first night I slept in the open bush in this country: it was in the Black Forest. I

woke about daybreak, after a confused sleep, and for
some minutes I could not fancy where I was, such
were the extraordinary sounds that greeted my ears:
the fiendish laugh of the jackass; the clear, flute-like note
of the magpie; the hoarse cackle of the wattle-birds;
the jargon of flocks of leatherheads; and the screaming
of thousands of parrots, as they dashed through the
forest, all joining in chorus, formed one of the most
extraordinary concerts I ever heard, and seemed at the
moment to have been got up for the purpose of wel-
coming the stranger to this land of wonders on that
eventful morning. I have heard it hundreds of times
since, but never with the same feelings that I listened
to it then.

The laughing jackass is the bushman's clock, and
being by no means shy, of a companionable nature, a
constant attendant about the bush-tent, and a destroyer
of snakes, is regarded, like the robin at home, as a sacred
bird in the Australian forests. It is an uncouth-looking
bird, a huge species of land kingfisher, nearly the size of
a crow, of a rich chestnut-brown and dirty white colour,
the wings slightly chequered with light blue, after the
manner of the British jay; the tail-feathers long, rather
pointed, and barred with brown. It has the foot of the
kingfisher, a very formidable, long, pointed beak, and a
large mouth; it has also a kind of crest, which it erects
when angry or frightened; and this gives it a very ferocious appearance. It is a common bird in all the forests
throughout the year; bred in a hole in a tree, and the

eggs were white; generally seen in pairs, and by no means shy: their principal food appeared to be small reptiles, grubs, and caterpillars. As I said before, it destroys snakes. I never but once saw them at this game: a pair of jackasses had disabled a carpet-snake under an old gum-tree, and they sat on a dead branch above it, every now and then darting down and pecking it, and by their antics and chattering seemed to consider it a capital joke. I can't say whether they ate the snake,—I fancy not; at least, the only reptiles I ever found in their stomachs have been small lizards. The first sight that struck me on landing in London was a poor old laughing jackass moped up in a cage, in Ratcliffe Highway: I never saw a more miserable, woebegone object; I quite pitied my poor old friend, as he sat dejected on his perch; and the thought struck me at the time that we were probably neither of us benefited in changing the quiet freedom of the bush for the noise and bustle of the modern Babylon.

There is a smaller species, the *Sacred Kingfisher*, which we used to call the Van Diemen's Land jackass: this is a real land kingfisher, nearly the size of a starling at home; bright blue above, light chestnut breast, which is much deeper in the male than the female, and white belly. This bird was sparingly dispersed over the bush, always seen in pairs; generally about the old gum-trees, in moist situations, by creeks or swamps. It bred in the hole of a gum-tree, and the old birds were always close to the nest. It has a shrill call-note, not unlike that

K

of the European wryneck, and was a summer migrant to our parts.

The real Australian *Kingfisher* is smaller than its European namesake, which it resembles much in habits and appearance ; it was, however, of a uniform purple-blue colour, and the breast was deep orange; it was a summer migrant to us, and a pair or two might be then seen on every creek : they bred in the hole of a bank, and the eggs exactly resembled those of the British bird.

No bush-bird to my fancy had a clearer or richer note than the *Magpie :* one of the earliest birds of morning, it was also one of the latest at night ; and the deep, flute-like evening song of the magpie was heard in the forest long after all the other birds of day had retired to roost. The Australian magpie is more like a rook in shape than its British namesake, but not so large and clumsy, and it wants the long bronze-tinted tail of the European bird ; it is, however, a graceful, elegant bird, and the rich black plumage of the breast and wings contrasts finely with the pure white of the back. The females and young birds are mottled grayish blue and white ; but I always fancied we had two species, the one mottled, the other black and white. The magpie is a very common bird throughout the bush during the whole year, often in small companies ; and in the autumn the old and young birds congregate in flocks. It is by no means shy, and one of the best cage-birds in the colony ; for they are easily tamed, and soon learn to imitate any call or noise. Unlike the magpie at home, it builds a careless shallow nest, and the female lays three greenish mottled eggs: the young magpies are excellent eating: the eye is brown-red.

I killed one single specimen of the *Black-backed Magpie* here, which is, I believe, common on the Sydney side, but had not been noticed before in this district: it might, however, have been overlooked; for it exactly resembled the common magpie in every respect, except that the back is black between the wings, instead of white.

We had another bird, which we called the *Black Magpie*, but which was a species of chough; about the size of the common magpie, but more slender; higher in the leg and longer in the tail; the whole colour sooty-black, with white wing-feathers, a long tail, and long, thin, dark, curved beak, like the British chough; the wing, when spread out, was very round, and the white pinions gave the bird a pretty appearance when flying. It was very common in our forests throughout the year, principally frequenting the large gums; was generally seen in small flocks, chasing one another from branch to branch : its call-note was a clear, soft, loud whistle.

The bird that we called the *Blue Jay* resembled its British namesake in no one particular. As large as a crow, very loose-feathered, the whole plumage one uniform dun-blue, with a yellow eye and large beak. It was common in our forests throughout the year, and the call-note was a loud whistle.

We had a smaller species of this bird about one third

less, exactly the same in habits as the other, but darker
in plumage and much rarer.

The *Swamp Magpie*, or mourning-bird, so called from
its black and white plumage, is an elegant little bird,
rather larger than the double thrush at home. It was a
winter migrant to us, and I generally found them in
small flocks in the belts of timber bordering the plains,
or on the edges of the swamps themselves, but scarcely
ever in the open, and almost always on the ground.
They were always shy and difficult to approach. The
plumage is rich glossy black and white, very strongly
marked, the beak and eye white. Their call-note was a
deep loud whistle, which I often used to hear long after
sundown.

The large *Carrion Crow* was common in our forests
throughout the year, but we saw most of them during
the autumn and winter. I think a few pairs bred with
us. It is larger than the British crow, being intermediate in size between that bird and the raven, which it
much resembled in appearance and habits. The whole
plumage glossy purple-black, the tail rather cuneiform. I
always fancied we had two varieties, the one smaller than
the other; and this in habits more resembled the British
rook, seemed to go in larger flocks than the other, and
in autumn congregated much on the wet plains. I never
saw a real rook out here. Unlike its British namesake,
which leads a solitary and persecuted life, the Australian
crow is rather a companionable bird than otherwise;
generally seen in small flocks, and often close to the

habitations of man. Its croak is loud but soft, and at times prettily modulated. They were our constant companions out kangarooing, and would follow us through the forests like sutlers on the skirts of a pursuing army, and at night, when skinning the dead kangaroo previous to bringing them home, the old carrion crows would perch themselves on a gum-tree above our heads and sit watching us till their turn came. Like

> " Raven on the blasted oak,
> Who waiting while the deer is broke,
> His morsel claims with sullen croak."

We had another species, rather smaller than the carrion crow, which it otherwise much resembled in shape, plumage, and habits, but the eye was clear bluish-white. We called it the *White-eyed Crow*. It was rather a local bird, generally seen in pairs, occasionally joined the other, but was nowhere very common with us.

One of the noisiest and most restless of all the bush-birds is the *Mocking-bird*, as we called it, for what reason I know not, as I never heard it utter any other than one note,—a long continuous hoarse cackle; and this was never still. It was about the size of a thrush; the upper plumage chestnut-brown, the under parts dirty-white, a bluish-white eye, and a long curved beak. They were not very common; generally in small flocks, in secluded places among the honeysuckles and shey oaks; continually in motion, chasing one another from tree to tree with a very sharp flight, all the while keeping up their peculiar hoarse call-note. They bred with us, but appeared far more common in the winter than at any other time.

The *Cuckoo* was another summer migrant to us, and of this we had three varieties,—the large gray cuckoo, the common cuckoo, and the little bronze or zebra cuckoo; and of this last I fancy we had also two distinct varieties, the one rather larger and much brighter in plumage than the other. Neither of the three had the call-note peculiar to the home bird. The large cuckoo was the rarest with us, but seemed to come the earliest. An odd pair or so (for all the species flew in pairs) were generally to be seen in the forest on any summer's day, flying about the tops of the high stringy-bark gums. All the three species had the peculiarities of the British bird in shape, beak, feet, and flight; and any one at a glance could tell to what class they belonged. The *Large Cuckoo* is half as large again as the common cuckoo, of a dull ash-gray colour, with a long pointed barred tail. It had a loud single call-note or whistle, often repeated when flying from tree to tree. The *Common Cuckoo* rather resembles its British namesake in colour, habits, and appearance; but the sexes do not appear to differ in plumage. The note was a simple call. It was more common than the last, and frequented smaller trees, such as shey oak and honeysuckles, whereas the large bird was always to be found among the high gums. The *Bronze Cuckoo* was a beautiful little bird, scarcely so large as the wryneck at home, the whole upper plumage and wings green-bronze, breast dull white, striped or striated with black; the under tail-coverts orange, and the tail- and wing-feathers barred with black. It has a very loud call-note for the size of the bird, rather resembling that of the wryneck; was, I think, the commonest of the whole species, and frequented small scrub, particularly small honeysuckles. Early in the season I used to find them much in the heather and low scrub; and I fancy they breed in the nests of the small brown wren; at least I once shot a female bronze cuckoo flying from such a nest, in which was a large spotted egg; and on dissecting the cuckoo, I found a similar egg inside it, but unfortunately broken. I can say nothing with certainty respecting the breeding habits of the other two, except that you rarely see either in small bushes, and during the breeding season I observed that the large cuckoo used to keep always about the same gum-trees.

Another summer migrant to our district was the *Summer-bird*, about the size of the jay at home, but more slender, of a slate-colour above, white under, with a black moustache, large black bill and legs, and full black eye. It was not a rare bird, always seen in pairs among the large timber, continually on the wing; and the call-note was one long soft whistle, often repeated while in the air. They flew with a slow dipping kind of flight, and soared over the tree-tops.

By far the commonest and boldest bird in the Australian forests is the *Miner*, or *Soldier-bird*, which, like too many of the human race, appeared to mind everybody's business but its own. Like the common sparrow at home, the miner was seen in all places and at all seasons, and, like that bird, was a "household word" with us. Always bustling about, on the broad look-out, let a strange bird but show itself, and a dozen miners, like so many policemen, were round it in an instant to drive it off. If the shooter is creeping quietly through the wood for a safe shot, it is ten chances to one that a miner spies him, and warns the prey of his approach; and if by chance a snake or stump-lizard shows a head, a congregation of miners will soon gather round it, and spread the news through the whole neighbourhood. They certainly are the most pugnacious birds I ever saw; and if they can't find any stranger to have a turn-up with, generally manage to get up an Irish row among themselves. The very snake-like head and well-guarded eye of this bird, and sharp beak, have quite a pugilistic cut. They are never still,—here, there, and everywhere, chattering, whistling, and chasing each other about from tree to tree. There was, however, something to my fancy very jolly in the habits of this bold bustling bird, and I used to fancy that those which frequented our tent knew me, and welcomed me as an old friend whenever I came home. The miner is about the size of the English blackbird, of a uniform light ashen-gray colour, many of the feathers edged with yellow, sharp beak and claws, bright piercing eye, and a yellow cere between the eye and the beak. They are common in all parts of the bush throughout the year. The note is composed of whistling and chattering, like a flock of starlings at home before going to roost. The flesh is bitter to the taste, like that of the starling.

We had three species of *Thrush*, two of them summer migrants,—the *Green Thrush* and the *Mountain Thrush*, and the common *Gray Thrush*, which remained with us throughout the year.

One of the sweetest sounds in the Victorian forest, to my ear, was the loud monotonous note of the green thrush, from the topmost branch of a high gum-tree, on one of those clear delicious mornings so peculiar to the Australian spring. Although not to be compared to the rich and varied song of the British thrush, there is a gush of melody in the few notes of the Australian bird equal to any of our finest songsters ; and as I have often and often stood at my tent about sunrise and listened to its wild desultory carol, borne upon the early breeze, laden with the fragrance of many a thousand blossoms, I have thought how dull and senseless must that block-head have been who described Australia as a land where the flowers have no scent and the birds no song. The green thrush is a fine bold-looking bird, about the size of the double-thrush at home, of a pale yellowish-green colour above, the under parts white, spotted with black, and a reddish eye. It builds a very pretty pendent nest, between two small twigs, and lays three large handsome mottled eggs ; in fact, I think the nest and egg of the green thrush prettier than any I ever took in this country. It was sparingly dispersed in pairs over the whole bush, but nowhere very common.

The common *Gray Thrush* is a dull-looking bird, of a uniform ash-gray colour, and in size and habits much resembling the blackbird at home. It was shy, kept as much out of sight as possible, and was generally seen feeding on the ground. It appeared to be the commonest of the three, and remained with us throughout the winter.

The *Mountain Thrush* of Australia is identical with "White's thrush" of Britain; and this thrush must have as wide a geographical range as any bird in the world, for specimens have been killed as far north as Sweden. It was by no means rare with us in the breeding season, being partially scattered in pairs over the tea-tree and other thick scrub. It is one of the shiest birds I know, and not often seen, for they frequent the thickest scrub, are almost always running on the ground, and rarely rise on the wing. I never heard it utter a single note. We saw them very rarely, and late in the season. I have killed them in the beginning of April, and taken their nest the first week in August, and I am not certain that some of them did not stop in our thick scrub throughout the year. The colour is uniform rich brown, the breast and belly light, each feather tipped or spotted with black in the shape of a crescent. It is about the size of the redwing at home. The nest and eggs very much resemble those of the British blackbird. The nest is very large, lined with coarse grass and fibres, placed at different heights in the tea-tree scrub, or on the large limb of an old honeysuckle. The eggs are three in number; in fact, all the thrushes, and many of the common bush-birds of Australia, lay but three eggs.

The *Wattle-bird* is a fine-looking bird, about the size of the British fieldfare, but longer; the general colour ash-brown, marked with black, a long thin pointed tail, a bright yellow tinge on the belly, and a red fleshy wart or excrescence hanging down from each ear; a powerful long pointed beak, and long sharp claws. This is one of the honey-eaters, which class of birds is characterized by a long horny tongue, feathered and fringed towards the end with fibres, for the purpose of gathering the honey and pollen from the blossoms of the trees. The wattle-bird remained in our forests throughout the year; bred in the small honeysuckles and shey oaks; the nest like that of the blackbird; eggs large, three in number, deeply spotted with red. It was found among the honey-suckles and gum-trees, in those particular seasons when the blossoms yielded the honey. In flight it much resembled the fieldfare at home. It has a very loud hoarse note or cackle, which we used to compare to the words " up with the rag," often and quickly repeated. In the end of autumn the old and young birds congregate in large flocks. They are excellent eating.

The *Australian Redwing*, as we call it, is another of the honey-eaters; not so large as the wattle-bird, one uniform greenish-brown colour marked with black, the under parts of the wing chestnut-red, the eye bluish-white. It was rather a shy bird, not so common as the wattle-birds; frequented the same localities, and bred in company with them on the small shey oaks and honey-suckles. The eggs were three, lightly spotted with red.

One of the bush wonders is the *Leatherhead*, or bald-headed friar, a curious-looking bird; not so large as the wattle-bird, of one uniform dun-blue colour, with black pencillings, a dirty-white breast and belly, white under the tail, which was long and square. The greatest pecu-liarities in this bird, however, are the head and neck, which, instead of being feathered, are covered with a thin black skin. The beak is large, with a sharp curve; and a high ridge or comb runs along the top of the head. It has a ruff or fringe of long pointed feathers, like a cock's hackles, at the bottom of the neck; the eye is reddish, bright, and deeply shielded, and the head and neck give the bird rather the appearance of a small vulture; and had it been larger, one might have supposed that it was a snake-killer; but for what purpose the head of so small a bird is so securely guarded I never could imagine. It is not a carrion-feeder, for the long feathery tongue proves it a honey-eater. They were very gregarious, building in small colonies. Large flocks used to visit our parts at irregular periods, and they then frequented the high gum-trees. I generally saw them in the middle of summer. They did not breed with us, but I remember seeing their nests in the small shrubs in a paddock under the Dandenong ranges. The leatherhead has the most curious and varied call-notes,—they can hardly be called a song, which it would be impossible to describe with the pen; a jargon of whistling, chattering, and cackling, which can be heard nowhere but in an Australian forest.

The *Wood Swallow* was another summer migrant to us; and of this bird we had two, if not three, distinct species common in peculiar localities. Both varieties used to associate, were gregarious in breeding, generally frequented the small open honeysuckle and shey-oak scrub on the edges of the swamps and plains, and I never met with them in the deep forests. The wood swallow is nearly as large as the British starling. The general colour of the common variety is dun-blue, light underneath, with a white eyebrow; the tail is fan-shaped, the middle feathers pointed, longer than the others, which is very apparent when the bird is on the wing. We used sometimes to kill this bird without the white eyebrow, and I fancied there were two species. The other variety was finer-shaped, and the breast and belly were brick-dust red. They were constantly on the wing, hovering over and dipping down on to the old honeysuckles where they bred. The nest and eggs resembled those of the European shrike. In autumn they congregated like the starlings at home, in great flocks on the low meadow-land, previous to leaving. The note was a twittering kind of call. I consider the wood swallow is very appropriately named.

One of the commonest of the small bush-birds throughout the year was the *Honey-bird*, or *Honey-eater*, and was met with all over the bush, among the honeysuckle-trees and flowering scrub. We had two species: the

largest, which was by far the commonest, was about the
size of the British yellowhammer, but longer, and not so
thick; it was a bold-looking bird, the ground-plumage
black, brown, and white; a white moustache on each
cheek, a white eye, long curved beak, and feathery
tongue; the wing-feathers edged with bright yellow.
This was a noisy, restless bird, had a shrill, loud call-note,
used to congregate in small flocks, and they often had a
battle-royal among themselves, which much reminded me
of many a similar scene with the old sparrows in the
hedge-rows of the stackyards at home. The other variety
was smaller, much duller in plumage, wanted the moustache, and had a kind of dark brown crescent on the
breast: it was rarer, and frequented more secluded
localities, such as deep gullies and thick tea-tree scrub.

I have seen a species of *Bee-eater*, which was killed up
the Plenty, exactly resembling the British bee-eater in
shape, but not so large or pretty; I never met with one
in our parts.

The *Warty-faced Honey-eater* is a very pretty bird,
nearly as large as the English starling, of a deep black
colour, spotted with bright yellow; a pale red naked cere
round the eye, and beak covered with small warts: it was
a rare and uncertain visitant to our district; generally
came in large flocks; flew high over the tree-tops, into
which they would drop, after the manner of the waxwing
on the rowan trees of northern Europe: they were wild
and shy, and the call-note was a low soft whistle.

The thirsty traveller, when wandering over these

parched and arid plains in the summer, gladly hears the "ching-ching" of the *Bell-bird* from the tea-tree scrub; for this is a sure and welcome omen that water is at hand. One of the greatest drawbacks from the pleasure of travelling through this country in the hot weather is the want of water. I have often walked for hours under a burning summer's sun without coming to a creek or water-hole; and of all the pangs to endure, those of thirst are the most intolerable. A man need never starve in the bush, but I have no doubt many have died for want of water here. Most of the creeks and water-holes lie so hidden in scrub and timber, that they are often passed by unnoticed; and often when we do come upon water, it is thick and muddy, and lukewarm from the rays of the sun. "But those who are parched with thirst do not stop to analyze the water. In tropical countries there is always tropical zest as well as tropical flavour." I always carried one of Hall's empty pound powder-canisters in my pocket, which I filled when I came to clear water; and from their shape and size these make the best water-flasks I know.

The bell-bird is about the size of the honey-eater, but much stouter made; the beak is very thick and powerful, and there is a red cere between it and the eye, which is red: the whole plumage is uniform greenish yellow. It is nowhere a very common bird, and is always met with in small colonies, in secluded places, by the side of creeks or water-holes, where large gum-trees are growing, from which they are continually flying up and down into the tea-tree, all the while uttering their loud, monotonous call-note. They breed in the tea-tree scrub, in company; the nest shallow, the eggs three, reddish white.

We had two species of *Swift*, which visited us at irregular periods during the summer; the one the large spine-tailed swift, and the other a smaller variety, which rather resembles the swift at home.

About Christmas, especially on a clear hot morning, a large flock of the spine-tailed swift would pay us a visit, stop for a day or two, disappear, and we, perhaps, should not see them again for ten days. Always in motion, hawking high in the air, screaming in wild joy, or dashing by us on the plains with the speed of an arrow, this is certainly one of the swiftest-flying birds in the world; about the size of a starling in the body, but in the shape of a pear; the wings very long and pointed, and the tail-feathers have each a sharp spine or prickle protruding from the end: the body-colour is sooty-black, the back and rump brocoli-brown, white towards the tail. The other species is much smaller, more resembling its British namesake; but the tail is square (without spikes), and the rump is white. The two species did not appear to associate much, and generally came to us in flocks by themselves on different days. We rarely saw either before the middle of December or after the end of March. I have heard that the spine-tailed swift breeds on the Heads and on some of the islands in Western-port Bay.

No two birds in Australia remind the emigrant of his village home in the old country so much as the *Swallow* and the *Martin*. There is a marked resemblance between many of the British species, and their namesakes in this country; but here we have the very birds themselves, hawking over the creeks and plains, and forming their clay nests under the shingles of the bush hut, just as we were wont to see them skimming over the meadows and rivers at home, " from morn till dewy eve," or building under the eaves of the straw-thatched cottage in the village streets. We miss, however, the pretty artless twitter peculiar to the British bird, for the Australian swallow has no song. Although the severity of the winter in these climes is so little felt that we scarcely notice the advent of the summer migrants with the same joyous feelings that we did at home, still the first sight of this elegant and cheerful little bird cannot fail to bring back pleasing recollections to the minds of all, for of all birds in every clime, the swallow is, perhaps, one of man's most constant and faithful companions. In colour and habits both birds in all respects resemble their namesakes at home, but they appear to be a little smaller, and I have often observed both building in large decayed trees by the side of the swamps. The nest is formed of clay, lined with feathers, often of the most gorgeous colours. I never observed a Sand-marten out here.

Strange to say, I never met with a single Woodpecker in this country, which would appear so peculiarly adapted to the habits of that class of birds, abounding as it does with such extensive forests, the old dead trees of which must afford shelter to millions of insects. We had two species of *Creeper*, the one much larger than the other. The large variety was rather a pretty bird, with a spot of chestnut-red upon each cheek. In habits they much resembled the British creeper; but the absence of the long thin bill peculiar to that bird, led me to consider them as more closely allied to the Nuthatch than the real Tree Creeper.

There is no true Skylark indigenous to Australia, but larks have been imported from England, and turned out wild. It will be a cheering sound in the ear of that man who has but lately left his English home, the clear shrill note of the Skylark in this land, where no single bird has any one long-continued song. And as cultivation increases, the country will gradually become more adapted to the habits of the lark. Nowhere are British cage-birds more highly prized than in Australia, and the simple carol of one of our commonest home songsters, when heard in a foreign land, cannot fail to raise pleasurable emotions even in the rudest and most untutored mind, for it speaks a language of youth and home familiar to all. We had a large species of lark on the plains, something between the bunting and the real lark, which we called the *Mounting Lark*. It was a very fine bold-looking bird, much larger than the common bunting, with the long powerful legs and claws peculiar to that bird; but the beak was large, and in shape resembled that of the lark. It was of a dark-brown colour, with black cheeks and breast, frequented the dry open plains, would run along the ground, rise high in air, drop and rise again, all the while uttering a loud wild carol, which, without possessing the melody of the European skylark, was a deep, rich, although monotonous, song. It was known among the shooters by the significant name of " Captain Flash;" was a summer migrant to our parts, as well as a smaller species of lark of a lighter brown colour, which was also found on the plains, and appeared to be a link between the lark and the piper.

The *Piper* was very common on all the dry plains during the summer, and resembled the Meadow Piper at home in appearance, habits, and call-note, but was lighter in plumage. Used to kill a large variety on the dead seaweed along the coast, which I considered the Rock Piper.

Five species of *Robin* were more or less common to the districts in which I have camped. The large black-and-white robin, which we also called the Magpie Sparrow, was the largest of all : a thick bird, larger than the Yellowhammer at home, pied black and white ; a summer migrant to us, and generally seen in pairs very sparingly dispersed about the small belts of honeysuckle on the edges of the plains. I never fancied this a true robin. The common Australian robin is smaller than its British namesake. The body-colour deep black, with a white forehead and dull-red breast. Of this bird we had three other varieties, the one a little larger than the last, the red on the breast much brighter and much more of it. This was often on the plains, the other more in small timber. We had another variety with a red forehead, which was not common with us, and very local. But the rarest of all was the purple-breasted robin, the smallest of all,—sooty black, with no white on the forehead, and a deep plum-coloured breast. This was a shy solitary bird, and I always found it singly in the thickest scrub. In habits the Australian robin resembles the home bird, but it has no song.

We had a curious little bird which we called the *Swallow diceum*, in size and habits much resembling the golden-crested wren of Europe. The body colour purple-blue, like the swallow, with a red throat and under-tail coverts. It was an irregular visitant to our parts, had a deep loud call-note, and frequented the large gums, being very partial to the bunches of mistletoe which grow on those trees; they are extremely difficult to see, on account of their small size and habits.

Of the *Wrens* we had about four varieties. The *Superb Warbler*, or blue wren, one of the most splendid little birds in the colony. The *Emu*, or pheasant wren, the smallest and most curious of all the bush birds; and two other species, but I am not certain whether these were true wrens, although we called them so. These latter we always found in small bushes by the edges of the creeks or swamps; they were both mottled, black, and brown, and one had a faint but rather pretty whistle.

The *Superb Warbler* is certainly rightly named, for I don't think there is a handsomer warbler in the world. This is a small bird, with a jet-black body, long fan-shaped tail; the neck, shoulders, and part of the back being covered with a little cape of long feathers of the most splendid bright ultra-marine colour. The blue wren is common at all seasons throughout the whole bush, frequenting small scrub and old honeysuckles, and is very partial to tea-tree scrub by the side of creeks. The male has a pretty little song, which he trills out when perched upon an old dead log, with his family round him,—for we rarely saw a blue bird without four or five brown-coloured birds in his company. The females and young birds of the year are plain dull-brown, with a light-blue tail, and some have a reddish throat. I do not think the male birds come to their full plumage till after two or three moults, and, like all the other handsome birds here, they are in best and hardest feather in the winter.

The little *Emu* or *Pheasant Wren* was the smallest bird in our parts,—scarcely larger in the body than a great bumble-bee. The whole colour is light-brown, the feathers loose and long, and the male has a pale-blue throat. The tail is about three times as long as the body, composed of six feathers (the middle ones much the longest) all clothed with fibres, after the manner of the tail-feathers of the native pheasant. It has very small wings, and weak powers of flight,—in fact, when flying it appears to have a difficulty in bearing its long tail. It is a busy little bird, and I liked much to watch a family of them creeping about the small scrub and heather like so many little field-mice. We generally found them in small colonies or families, among heather, low scrub, or long grass on the plains and swamps: they were very hard to rise, and when on the wing easily knocked down with a small bush or cap. The male has a weak but pretty little song.

I have seen two or three species of *Sedge Warbler* in the reeds by the side of the creeks and swamps, and one used to keep up a continuous little song throughout the summer nights, not unlike that of the sedge bird at home.

The *Satin* or *Shiny Bower Bird* was a rare and only an occasional visitant to us, generally appearing in the autumn and winter, and those which we saw in our district were principally the yellowish-green birds, sometimes accompanied by an old black cock. The old male satin bird is a splendid bird, nearly as large as the jay at home; the whole plumage a beautiful deep-purple glossy black, the eye bright-blue, the beak, which is long, thick white. The old males are very shy and very rare in proportion to the yellow birds. Like many of the parrots, the males do not come to their full purple plumage until after about the fourth moult; the standard colour of the females and young birds being greenish-yellow, mottled, which in the males becomes every year chequered with black, till they attain their full plumage. The note of the satin bird is a kind of loud guttural hiss. They are very common in some parts on the ranges, and they come down much into the bush gardens when the peach is in bud, and when the grapes are ripe, and are at such times very destructive. When they came into our parts the yellow birds were by no means shy; they frequented

the gum trees and tea-tree scrub. They breed in thick
tea-tree and other scrub, generally in gullies and near
the nest; the old birds form a sort of bower of dead
sticks, which they ornament with parrots' feathers, &c
If the old male is shot the female will soon find another
mate; and I have shot three cock birds from one bower
up in the Dandanong ranges. The flesh is rather bitter,
like that of the starling at home, and they are not much
fancied for the table of the epicure, but often found
their way into our bush larder.

We had two species of *Fantail*, the largest, which we
called the *Stock-Whip Bird*, or shepherd's companion, had
rather the appearance of the pied wagtail at home, but
was much larger and thicker, of a sooty black-and-white
colour, with a long spreading fan-shaped tail. It is a very
lively bird, always in motion, and its attitudes are very
elegant as it flits from tree to tree, or runs along the
ground with outspread tail, uttering a grating call-note,
something similar to the springing of an old watchman's
rattle, but of course not so loud, ending with a sharp
smack. It is common on the plains during the summer,
often among sheep, upon whose backs I have seen them
perched like the starling at home. They were generally
in pairs, bred in the belts of honeysuckle and shey-oak
on the edges of the plains, and the nest is very curiously
formed,—a small round cup stuck upon the bare surface
of a large limb, without any shelter, looking just like a
nob or wart growing to the bark. The other variety
was much more elegantly formed, also with a spreading

tail, shiny black-and-white in colour, and the throat and
chest were faintly tinged with salmon-red. This was
hardly so common a bird as the other, but frequented
much the same localities. The great difference between
the two birds lay in the shape of the beak and gape—
the beak of the larger species resembling that of the
swallow; the other was more like the night-jar. They
were both fly-catchers, but I do not think they were the
same species.

We had no real wagtails in this country, but I observed our common little pied wagtail very common in
South Africa, in the months of January and February.

The bird which we used to call the *Fly-catcher* was
much smaller and more common than either of the fan-
tails, which it, however, resembled in shape and habits,
and was pretty generally dispersed over the whole bush
throughout the year. It was of a light variegated brown
black-and-white colour, with a long spreading tail, and
principally frequented the honeysuckle scrub. Its note
was a kind of grating chatter,—loud for the size of the
bird. We had another variety, light chestnut-brown,
but this was very rare with us, and I once saw a specimen of this bird, pure white, but whether it was a distinct species I am unable to say.

Another little bird, which we called the *Pretty Fly-
catcher*, very much resembled the salmon-throated fantail,
but was much smaller, and the colour was more glossy.
It was a rare and solitary bird, and I generally used to
kill single examples in the thick scrub.

The *Great Shrike,* or cobbler's bird, as we called it, was rather a common bird in our forests throughout the year. It is a real shrike, as large as a thrush, of a dirty-white and blackish-brown colour, very bold, and generally seen singly or in pairs. It is very bold, and one or two were always about the bodies of the kangaroo which hung near the tent. It has a loud clear whistle, and is, I believe, an excellent cage bird. We had a smaller species which we called the *Stringy-bark Shrike,* of a chestnut-and-white colour, which generally frequented the tea-tree scrub, and seemed more to resemble the thick-heads than the true shrikes: this was a summer migrant to us.

The *Thick-Head* is a species of oriole, peculiar to thick scrub; and of this we had at least two varieties. The one large as the bunting at home, of a uniform greyish-brown colour, the other much smaller, lighter in plumage, with a gray chin. The larger variety was rare with us. This bird derives its name, I suppose, from its thick chubby head.

We had two species of *Oriole,* as we called them—the one a little larger than the British yellowhammer, of a bright yellow-and-black colour, rare, and principally found in the tea-tree scrub. The other smaller and more common, of a dull yellow-and-gray colour.

There was a bird on the ranges which we called the *Crested Shrike,* in size and shape resembling the cross-bill at home, and the beak was as large and powerful, but not crossed. The body colour was yellowish-green,

with fine black and white markings, and a large black crest
on the head. In the male the throat is deep black, in
the female dull. It was not a very common bird; generally seen in pairs high up in the gum or stringy-bark
trees, and the call-note rather resembled the mewing of
a cat.

We had no real titmice in our districts.

The *Cobbler's Awl* bird was a pretty little bird of a
chestnut-brown colour, with white belly and a black
crescent round the breast. The beak was long, thin, and
curved, and the bird bore some slight resemblance to the
humming-bird, of which class we had no real varieties
here, the range of these little birds being confined to
within 40 degrees north and south of the equator. It
was common throughout the year in the small honeysuckle and other scrub, but was rather local.

Whilst watching in the thick tea-tree scrub by the
side of a creek for ducks, the ear is often startled by a
loud whistle ending in a sharp smack like the loud crack
of a whip, something like that of the stock-whip bird,
but a great deal louder. This is the call-note of the
Coach-Whip Bird, a large species of fly-catcher, nearly
the size of a thrush, of a uniform light cinnamon-brown
colour, with a long wedge-pointed tail and small round
wings. I do not believe the bird is so very rare in
favourite localities, but as it always keeps in the
thickest tea-tree scrub it is oftener heard than seen.
There was a smaller species, something resembling this
bird, but which had not the same loud note, and which

I generally found in low scrub, on the edges of the tea-tree.

We come now to the finches, and the members of this class are small in proportion to the soft-billed and honey-eating birds.

We had three varieties of the wax-billed finch, or blood-bird, as they are wrongly called in the bush, on account of their blood-red rumps, the real blood-bird being of a bright-red colour, and not met with in this district. The *Little Wax-Bill*, which was the smallest and commonest of all, being no larger than the liskin at home, of a deep-brown colour, a pointed black tail, a thick beak, red core round the eye, a bright-scarlet rump, and a red mark over each eye. This was a gregarious bird, and generally met with feeding in flocks, on the ground, among the honeysuckles. *The Guinea-Hen Finch* was larger than the last, but hardly so large as the linnet at home: of a dark-gray colour, striped and marked with black, a bright-red rump, a short dark tail, the feathers barred, like that of the British wren. This bird was usually seen in pairs, among the small shey-oaks and tea-tree scrub. These two species remained with us throughout the year. But by far the most elegant, and in our district the rarest of all, was the *Spotted-sided Finch*, a summer migrant to our parts, very similar in shape and size to the last, but of a pure white colour, with gray-and-black markings, six or eight deep-black spots on each side, a bright scarlet rump, and pale-red bill. This little bird was sparingly dispersed in pairs throughout the summer, over the honeysuckle and shey-oak scrub, where they bred, and in the autumn they congregated previous to leaving. The beak of all these birds is thick, of a reddish colour, having the appearance of being moulded in wax, whence their name. None of them had any song, but merely a call-note, or chirp.

We had also another little *Black-and-White Finch*, something in appearance resembling the last bird, but of a much duller colour; with no red on the rump or spots on the side, and the beak was dark. It was a very common little bird with us, used to congregate in large flocks on the plains and open meadow-land, flew in jerks, like the wagtail at home, and appeared to remain with us throughout the year.

However monotonous the call of the *Diamond Sparrow* may sound in the ears of that man who has always been accustomed to the rich melody of the various warblers that frequent the groves and thickets of Europe during the summer season, it brings with it a cheery welcome to the bushman, as the first notice of the arrival of the birds of summer into this part of Australia,—one of our earliest spring migrants. This elegant little bird frequents the large gums and honeysuckles; generally in open situations, rarely in the depths of the forest, among the branches and leaves of which it runs after the manner of the British titmouse, continually uttering its monotonous call-note, " Twit, twit, twit," loudly and quickly repeated. It is a pretty little bird, about the size of the guinea-hen finch; the upper plumage ash-

gray and white, barred with black and yellow, and three
or four small red spots on each wing. They bred in the
holes of the trees, and the eggs were small and white.
Some of our spring migrants appeared to come very
early, and I have noticed the swallow, the marten, and
the diamond sparrow the first week in August, but the
majority of them came to us early in September.

The last on our list of the small bush birds, and certainly one of the handsomest little birds in the colony, is
the *Diamond Bird*, which rather resembled the last in
shape, habits, and appearance, but was much smaller
and prettier. It would be in vain to attempt to do
justice, in a written description, to the varied and beautiful plumage of this handsome little bird. The general
colour, ash-grey and white, but spotted and spangled all
over with red, yellow, orange, and black, and the tail
coverts rich dark-red. It was very common in some
places among the large gums in the deep forests, and we
rarely found this bird and the diamond sparrow in the
same localities. The habits of the two are similar, and
the call-note of the diamond bird, although not so loud
and pretty as that of the diamond sparrow, is very loud
for the size of the bird. It bred in old logs, and sometimes in a hole in the ground. It was a summer migrant
to us, but I once saw a small flock in the winter.

Many of the birds above described bear a strong resemblance to their European namesakes, so much so, in
fact, that we can class them by their peculiarities in
shape and general appearance. But we now come to a

class only found in southern climates, and which for beauty of plumage, have no rivals in the old world—the *Parrots*, and I do not believe any country can be richer than this, certainly not in individuals, whatever it may be in varieties of this tribe of birds. At particular seasons they swarm over the whole bush. I do not know how many different species are met with throughout this country. We had about twenty kinds, more or less, common in our forests, and I have seen many other species, among them the cockatoo parrots, from other parts, which were strangers to us.

The cockatoos, on account of their size, stand first on the list.

The *Black Cockatoo*, or black toucan—for it has not the crest of the cockatoo,—is the largest of all this species. It is a fine bold-looking, but by no means handsome bird; the body full and round, larger than a crow; the tail long and spreading; the wings round when extended; the head large, the beak very powerful; and the old bird has a kind of crest which it can erect when angry or frightened, and which gives it a very ferocious appearance. The ground colour over all is deep-black, the feathers edged with yellow, which, as well as the spots on the tail, is much brighter in the young than in the old birds: the cheeks sulphur-yellow, and the tail-feathers spotted with the same colour. An old bird will measure about two foot from the beak to the end of the tail. The black cockatoo was common in our forests from about December, when the old and young birds came down from their breeding-places, and remained with us during the winter. They did not breed in our neighbourhood, but I think they went to nest very early, for I once shot a female in May with a large egg in her. They principally frequent the honeysuckles, but are often in the large gums. The old birds are very shy, and have a loud hoarse call-note or cackle. When they first come they are in large flocks, and they then always frequented the large honeysuckles, over the tops of which they would fly, or rather float through the air, with a wavering kind of flight, toying and playing with each other, after the manner of the rook at home. As the winter advanced they appeared to separate, and, although you hardly ever see a *single* bird, they disperse themselves much more generally over the forests. The young birds are excellent eating. Their principal food appeared to be large seeds and grubs, and they score the young honeysuckles round with their powerful beaks in search for these latter as if cut with a knife.

There is another variety of this bird, the tail-feather spotted with red. I only knew of two specimens being killed in our district, but I believe it is not uncommon near the Head.

I believe there is a third variety, the tail barred with red, which is very rare. This I never saw.

The *White Cockatoo* is very common in many parts, where they congregate in immense flocks; and I have seen a large patch of meadow ground covered with them like a sheet of snow. They were comparatively rare in

our districts, and I never saw them in large flocks; but
an odd pair or two used yearly to breed in our forests.
The white cockatoo is a handsome bird, as large in the
body as the last, but the tail is very short, and it has,
consequently, a much rounder and thicker appearance,
especially when on the wing. The whole body is pure
white, and the crest, which is very long, sulphur-yellow,
and the wings are tinged with yellow underneath. It is
a very wary shy bird, and the call-note nothing more
than a loud hoarse scream. Although apparently wilder
in the bush than the black cockatoo, it is much oftener
seen as a cage bird. They are excellent eating, and
when stuffed and roasted in the same way, can hardly
be known from a duck. I recollect we used to cook
wood-pigeons at home so, and, when eaten with the fenman's duck sauce, a little port wine, cayenne pepper,
and a slice of lemon, we could not tell them from
widgeon.

There is a variety of the white cockatoo with rose-
coloured crest, but this I never saw here.

The most curious looking of all the species is the
Yan Kate, a bird nearly as large as the white cockatoo,
of a dirty-white colour, the shafts and under parts of the
feathers, and the down rose-coloured. It had no crest,
and the beak was not like that of the other cockatoos,
for the upper mandible projected with a kind of thin
hook, more than an inch long, over the under one, which,
with the large bare cere round the eye, gives the bird a
most grotesque old-fashioned appearance. It was not

very common, and I generally used to find them in pairs among the stringy-bark trees, but I have occasionally seen them in small flocks. They are very shy, and it is difficult to get within gunshot of them. When disturbed, they would take a long flight round and round, making the forest re-echo with their loud call-note, "kakadua, kakadua," often and quickly repeated.

We had a species of small cockatoo, which we called the *Corella ;* the body was grey, tinted with yellowish green, and the male had a long, thin, crimson crest. In the female, the crest is yellow. It was not quite so large as the African grey parrot, which it much resembled in shape. They were only occasional visitants to our parts, and I always saw them in pairs.

By far the finest parrot that I have seen in Australia, is the *King Parrot*, which was, however, very rare in our forests, and what few I killed were principally immature birds, for the king parrot, like the satin-bird and the scarlet lowry, does not attain its full plumage until after the third or fourth moult. Although rare with us in certain places they are as common as the red lowry, which bird it seemed, as far as I could see, to resemble much in habit, and I think they are of the same species. The king parrot is not much less than the magpie at home. The plumage of an old male is a dark green body, with flaming red breast. The females and young birds are much duller in plumage.

Although not so large a bird, I think an old male *Scarlet Lowry* quite as splendid in plumage as the king parrot. It is next in size to this bird; the whole colour of a gorgeous deep red, the feathers edged with black, and the wing-feathers and tail have a dark purple shade. It is a magnificent parrot, and, as they float through the forest, they strike the eye like a flame of fire. The male and female are alike in plumage, but the female is smaller, and the tints much duller. For the first two or three seasons, the immature birds are greenish yellow, and they then go in flocks, and feed much upon the ground about the homestead, in company with the rosella. These are called the "green lowry," and they were the tamest of all the parrots. The red lowry was by no means rare in our forests, pretty generally dispersed over all, and they much frequented the gum-trees and scrub near water. All the parrots draw much down to the creeks and water-holes, where they are very fond of washing. Sometimes the scarlet lowry are seen in small flocks; sometimes, but rarely, they associate with the green, but we generally see them in pairs, male and female, by themselves. It is by no means a shy bird, and feeds principally on seeds.

The *Rosella*, when full plumaged, is a handsome bird, and is known in England better than any of the others, as the common cage-parrot from this country. The ground-colour is green, prettily variegated with red, yellow, and white; the head and throat crimson, and it is nearly as large as the lowry. It was very common throughout the whole bush, in particular localities. Unlike the other parrots, I do not think the rosellas migrate

much, but keep about the localities where they are bred.
We generally saw them in small flocks, and they were
sure to be about the little bush-farms. It much resembles the lowry in shape, habits, and flight; but, unlike that bird, is often seen feeding on the ground; and
it is altogether a tamer and more domestic kind of bird.
The male and female are alike in plumage, but the male
is much the handsomest.

One of the most dashing of all the parrots is the *Blue
Mountaineer*, which, unlike the three last, is a honey-eater.
On this account, the blue mountain parrots are certain
migrants to and from different districts, and their migrations are regulated by the state of the blossoms of the
gum and honeysuckles, upon which they feed; not that
they ever entirely left our forests, for I rarely went out
at any time without seeing a pair or so. But the large
flocks of them only come at such times as the trees are
full of honey, and depart as suddenly as they come.
The blue mountaineer is a splendid parrot: body sea-
green, head lavender-blue, the breast beautifully mottled
and watered with red, yellow, and orange; the tail green
and sharp pointed, the middle feathers longest, but
altogether much shorter than in the other species; the
wing-feathers with each a large spot of yellow, and the
under-wing coverts flaming red, which gives the birds a
splendid appearance as they dash through the forests, like
lightning, screaming as they go in all the wild joy of their
native freedom. They are always in larger or smaller
flocks, do not associate with the other parrots, and are never

seen feeding on the ground. Although each one is different after its kind, it would be hard to say which was the handsomest bird, if an old full-plumaged male of each of these four parrots were laid before one on a table. They are capital eating, and, as they come in good flocks, are much sought after "for the pot." They have a loud, grating, hoarse scream when flying; and during their migrations I often used to see immense flocks pass over high in air; in fact, this and the little green paroqueet seem to fly higher than any other birds in the colony.

We had a curious ground parrot, common in the long grass in the plains, on the heather, and often in low tea-tree scrub (sometimes up to the knees in water) called the *Swamp Parrot*. I have heard some very learned ornithologists call it the *Pheasant Cuckoo*, which I consider a very far-fetched name. The tail certainly is shaped like that of the common pheasant, and it is barred, and here the resemblance ends; but in what respect this bird resembles the cuckoo, I never could make out, seeing that it lives on the ground, has the beak of the tree-parrot, and the call-note is nothing more than a faint twitter. The swamp-parrot is an elegant bird, both in shape and plumage; nearly as large as the rosella, but not so plump. The ground colour, light sea-green; every feather of three colours, green, black, and yellow; a long pointed tail, the feathers barred with black and yellow, and a red forehead. The shape of the beak, head, and body, is that of the parrot. But the legs are long and bare; the claws long, straight, and pointed. In fact, it is a tree-parrot with the foot of the lark. It lives on the ground (but I have seen them perch on the tea-tree scrub), runs much and quickly, is hard to rise, flies in jerks, goes away very sharp before a wind, and is very pretty shooting, rising from the grass and heather. We used to find them during the whole year, frequenting different localities at different times; and although they could scarcely be said to flock, I generally rose three or four on the same spot. Dogs will set them like quail.

Another splendid parrot is the *Green Leek*, and this was by far the rarest of all the species with us; for I only knew of one example being killed, and this was an old bird, on a dry sandy rise near a swamp. It is about the size of the last, which it rather resembled in shape; but the beak is larger, and the tail not so long. The body colour is dark leek-green, the head yellow, the throat and breast orange and yellow. I know very little of its habits, but I believe it is common on the stringy and iron bark ranges.

The commonest of all the paroqueets is the common *Green Paroqueet*, which in shape and habits rather resembled the blue mountaineer, with which bird they much associated. They were both honey-eaters, and their migrations were regulated by the same causes. Although in general much commoner than the blue mountaineer, at certain times they would disappear altogether from the forest, and then again come " not as single spies, but in battalions." This bird is about

half the size of the mountain parrot, of a bright green
colour, the back light brown, yellow shoulders, and red
under the tail, which is short and pointed. These and
the blue mountaineers were the only two species that
seemed to associate in flocks. It is a sharp-flying little
bird, has a shrill scream; generally frequented the
gums; and it was pretty to watch them, creeping like
mice among the bunches of blossoms, when feeding.
They were very rarely on the ground under the trees.
They are very plump, and excellent eating.

We had a smaller species of this paroqueet which
resembled it in all respects, but it was only half as large,
and much rarer. They used to associate.

All the parrots come into their best plumage about
May, remain so till December, when they begin to get
dull and ragged. The birds in this country appear to
moult but once in the year, just after the breeding
season, and are a long time before they come into good
plumage. No birds here are of much value to the
collector, from the beginning of December till the end of
April.

Owing to the dense foliage of the trees, the brilliant
plumage of the birds in the Australian forests does not
strike the eye, as might at first be supposed; and as is
the case with the small flowers, it is not until they are
collected and closely examined that one sees how beautiful they are.

The swift *Flying Loriqueet* was by far the handsomest
of our common paroqueets, and stands in the same rela-

tion to them as the blue mountaineer does to the parrots.
In fact, there is a resemblance between the two birds in
shape, flight, and habits. The loriqueet is smaller than
the green paroqueet, but much finer made; and the two
middle feathers of the tail are long and pointed, project-
ing beyond the others, which gives the bird a very sharp
appearance. The plumage is green, prettily marked with
blue and red; the shoulders are crimson, and the tail
faint rose coloured. It was an uncertain visitant to our
forests; seemed to be the most common in the beginning
of winter; but at irregular periods all through the year,
except in the very heart of the breeding season, large
flocks would come down into the timber. This bird is
rightly named; for, as they dash through the forest, they
fly almost with the speed of the spine-tailed swift. Both
this and the green paroqueet are honey-eaters.

Very few of the parrots breed with us. I have taken
the nest of the rosella out of a hollow tree; eggs three,
and white. I found the eggs of the swamp parrot, four,
white, and more oblong than those of the other species,
which are generally round, on the ground, among the
heather; and I took the eggs of one of the ground paroqueets out of a hollow tree; but I do not think any of the
others breed in our forests, except perhaps an odd pair or
so of mountaineers. I do not fancy any of the parrots
are gregarious in breeding, but that they breed in odd
pairs, generally dispersed over the forests.

Occasionally, but very rarely, a flock of the *Budgere
Gar*, or *Shell Paroqueet*, would pay us a visit; and I

recollect, in the middle of the summer 1854, our gum-trees swarmed with them. They stayed about a month, when they suddenly disappeared, and only an odd straggler or so has been since seen in our district. I should say that this is the handsomest paroqueet in the colony, and is well known at home, as a cage-bird, by the name of the zebra paroqueet. It is smaller than the loriqueet, which it rather resembles in shape. The ground colour is light sea-green, prettily striped, and variegated with yellow and black; a light yellow forehead, and three or four deep purple spots on each cheek. But it is impossible to do justice to the beauties of this class of birds in a pen-and-ink sketch. The shell paroqueet is very common on the Adelaide side.

We now come to the *Ground Paroqueets*, and these can be easily distinguished from the others by their long thin legs, straighter claws, and smaller beaks; and they can all run well upon the ground, which the other species cannot. We had, I think, two varieties of common ground paroqueets, which were always in flocks on the plains or in the heather, often under the large honey-suckles, and they appeared to remain with us throughout the year. Unlike the swamp parrot, these little birds fly much into the trees, although they always feed upon the ground. They are smaller than the loriqueet, of a light green above and bright yellow below; the tail long and pointed, yellow underneath. One variety had blue on the forehead. The other was a duller and plainer bird.

But the *Red-shouldered Ground Paroqueet* was by far

the prettiest of all, and this was a regular summer migrant to our parts, and generally seen in pairs or small flocks, of four to six in the forest, feeding on the ground, on dry sandy rises, under cherry-trees or small gums. The general plumage of this paroqueet is green and yellow, and it has a dark red spot upon each shoulder.

The lowan, the emu, the wild turkey, and the native pheasant have already been described.

My list of game birds includes the six species of quail, the pigeons, the snipe, plovers, native companion, coot, moorhen, dabchick, bittern, herons, white cranes, egret, nankeen crane, spoonbill, and ibis.

I have already described the large crake peculiar to Port Philip; but we had two smaller varieties, which, although they in habits much resembled the water-rail, were, in my opinion, true crakes, and I considered them identical with the spotted and lesser crakes of Europe. They both frequented the long grass by the edges of the lagoons and swamps, were local, but by no means rare; I think the lesser crake was the most common, and I rarely met with both in the same localities.

The dark variety much resembled the spotted crake of Europe in shape and plumage. It was larger than the other, the beak longer, and dark green, the eye vermillion, eye-lids and legs red; the head was chesnut red. I used to kill two varieties; in one the head was dark, but as it resembled the other in every respect otherwise, I fancied it was the same bird in a different state of plumage.

The lesser variety was lighter in plumage and smaller

than the lesser crake of Europe, and I should fancy must
be the smallest wader in the world; the beak was light
green, red at the root; the legs green, the eye reddish,
the eyelid dark.   I rarely saw either on the wing; they
would run among the tussocks of swamp-grass like mice,
and the dogs often chopped them.

I cannot say that I ever identified a true water-rail
out here.

A species of dunlin, or small stint, came in the autumn
on to our plains in large flocks, but I never saw them in
the breeding season.   It much resembled the British
dunlin, and I fancy we had two if not three varieties;
at any rate, the specimens I killed used to differ very
much in size and plumage.

Occasionally, but rarely, I have killed a bird in every
respect resembling the European long-legged stilt, with
a beautiful red eye.   I always found solitary examples,
generally standing in the shallow water at the edges of
the lagoons.

We had a species of large wader, which I fancied
rather resembled our bar-tailed godivit.   It was known
among the shooters by the name of the sea-snipe; light
grey and mottled.   I generally found them singly or
in pairs, both along the coast and in the marshes; it had a
loud, long, single call-note, which I often used to hear
after dark.

The smallest of all the stints with us was a little light-
plumaged bird, with a chestnut head; it was not nearly
so large as the English Kentish plover, and used to

congregate in the autumn in flocks, often on the sea-beach, and on the sandy shores of those lagoons that lay near the sea.

I now and then killed a large grey sandpiper, by itself, on the coast, nearly as large as the dotterel; I never saw them in flocks, and with us they were wild and shy, and always singly on the beach.

A small species of plover frequented the sandy margins of certain lagoons, and although I generally found single birds, I have met with them in small flocks. We called it the bull-eyed plover; it was a true plover, rather larger than the ring dotterel, which it resembled in call-notes and habits, and had a red arc round the eye.

I did not consider this district nearly so rich in numbers or varieties of the smaller waders as the wild marshy nature of the country would lead us to expect; nor were the coasts at all rich in sea-fowl; it is true I paid but little attention to the smaller water-birds, for when beating the swamps or up and down the sea-coast I was after larger game; but I generally noted every bird that I saw, and certainly here I met with much fewer birds than I used to fall in with when coast-shooting in England; doubtless there are many other species which I overlooked, and I should certainly recommend the Australian naturalist to pay attention to this class of birds; for no country in the world is more fitted to their habits, and both the crakes and some others of the smaller species which I killed appeared to be but little known here.

We had two species of avocet, as we used to call both; but one was not a true avocet, for though it resembled the real avocet in shape and in the peculiar formation of the feet, the bill was straight, and the body mantle was light red instead of black; the real avocet exactly resembled its British namesake in colour, but had not the large white spot on the wings peculiar to that bird. Both species used to come on to our coasts at uncertain times; used occasionally to associate; but I think the real avocet was the most common.

Of the *Grebes*, besides the dabchick, we now and then killed a large crested grebe in the bay, but this I never saw inland. I never myself saw the snake-necked grebe here, although I have seen a specimen in Melbourne, said to have been killed in the neighbourhood.

We had neither the merganser or goosander here that I know of.

The black swan, the two wild geese, and nine species of wild duck, have already been described in my list of game birds.

We now come to the gulls and terns, and of these birds again our coasts were rich neither in numbers nor varieties.

Of the gulls, I could only identify three species in our district: the great black-backed gull, the lesser black-backed gull, and the common, or, as we called it, the pigeon-gull, with a white eye; all these resembled their British namesakes in habits and appearance, and perhaps the pigeon-gull was the commonest. We had, however, a large, dark-grey mottled bird on our coast, which I used to take for the young of the great black-backed gull; this and the young of the common gull were excellent eating.

I am only certain about three terns: the large Caspian, the common, and the black terns; and these all appeared exactly like the European birds. I have killed another species, something like the roseate tern, and I have heard of a smaller tern, about half the size of the common tern, which I never killed. Both the common and Caspian tern used to frequent our coasts, and the large Caspian tern was by no means rare; I used to shoot the black tern on the swamps and plains, where they would come occasionally in large flocks, especially in the autumn; but I never saw them here in the summer, and our coasts afforded no suitable breeding places for the other species.

The *Oyster-catcher*, or Sea Magpie, as we called it, was by no means rare on these coasts. It was just as noisy and restless a bird as its British namesake, which it appeared to resemble in all respects.

At times, when a large shoal of small fry set into the bay, hundreds of *Gannet* would follow them, dashing down headlong into the water, exactly as we used to see them on the north coasts of Britain. The gannets here appeared to resemble the British bird in all respects, but I fancy the chestnut on the head was darker, and did not extend so far. The young birds of the year are mottled, after the manner of the young gannets at home. The eye of the mature bird is transparent yellow white. Once while walking along this beach, an immense bird, as large as an albatross, flew by me. I could not get a shot at it, and I never before or since saw so large a sea-bird on these coasts.

We had two species of *Cormorant*,—the large one more like the shag than the cormorant at home; but it had no crest, and the colour was not bright. The smaller species, which we called the diver, had a white breast, and frequented creeks and inland lagoons, often very far away from the sea-coast.

The *Curlew* was at times common on our coasts, generally in small flocks, but I never, in the breeding or at any other season, met with them inland in our district. Although there were many miles of barren moorlands in the neighbourhood of the beach, they had not the character of those wild moors on which I have seen this bird breeding at home. The Australian curlew is very like its European namesake. The call-note is exactly the same. And when the sea was going down after a heavy autumnal gale, which had driven the birds on shore, I used to enjoy a walk by the sea-side; for at such a time the long melancholy wail of the curlew, the shrill cry of the oyster-catcher, the loud hoarse bark of the large gull, all blending with the wild whistle of the wind, and the regular unbroken roll of the surf as it set in upon the shingly beach, would strike the ear like rich and varied music, although played upon nature's rudest and wildest chords.

I cannot say that ever I saw a true whimbrel out here.

The *Pelican* was not rare on the mud flats of either of these bays, but I never saw them on the sandy beach. It is the largest sea-bird on these coasts; the body colour pure white, the pinion feathers deep black; the beak long and broad, and the pouch large enough to hold a man's head. They are easily crept up to when sitting half asleep on a sea-bank, but they are of little value to the gunner, and it is therefore a pity to shoot them; for they give an interesting wildness to the scenery on these coasts, whether when passing at evening with motionless and expansive flight over the mud flats to their feeding-grounds, or when far out at sea they float buoyantly on the surface of the water, rising and falling with every billow.

We had a small species of penguin in our bays, which, although rarely to be seen on the water, was often washed up dead on the beach; it was not more than half the size of the king penguin, so common on the coast of Africa, which it exactly resembled in shape and texture of plumage; but the body colour was shiny blue above, white underneath.

The *Petrels* close my list; and of this bird we had, as far as I could see, three varieties. The *Mutton-bird*, which was the largest of all, nearly the size of a pigeon, of an uniform dark dun colour; a smaller variety as to blue and white; and the common little storm petrel, smallest of all. Strange to say, I never myself killed a

single variety of either on these coasts; and all the
specimens I found were washed up on the beach dead.
Often after a heavy westerly gale, I have picked up a
dozen of the small birds on the coast, and I never could
make out what killed them, unless they were beaten
down by the violence of the storm. The blue-winged
variety was rare, and I only met with two specimens of
the mutton-bird in our bay. But there are islands somewhere off the Heads where both these birds must breed
in immense quantities, for boat loads of the eggs of the
mutton-bird are brought sometimes in the season up to
Melbourne for sale, and dried mutton-birds are a staple
article of commerce on some of these coasts. I have seen
more than a hundred dozen of the blue variety brought
in dead by one fishing boat.

So much for the different birds that I met with out
here, and I do not believe many more species are to be
found in this district, although I have, no doubt, omitted
some. My list will give a pretty good general idea of
the ornithology of Port Phillip, and this is all that it pro-
fesses to do. This part of the country is certainly not a
first-rate station for a collection, as most of the prettiest
and rarest Australian birds, such as some of the parrots,
the rifleman, the regent-bird, and many others, are
strangers here. But the only way to obtain a knowledge
of the natural history of any country is to compare the
notes of naturalists kept in different districts, good or
bad ; and when one looks upon the map, and sees what
a mere speck the district which I have been describing

appears upon the face of such a land as Australia, it will
be easily seen that no one man, by his own unaided research, could ever obtain a knowledge of the ornithology
of this country. Small as was my limit, and barren as
it might have been when compared with other districts,
I was always finding something new; and I have no
doubt, were I to go over the same ground again, I
should fall in with very many things that I had overlooked; for as one of our best field naturalists—White,
of Selborne—well observes, " It is with zoology as in
botany; all nature is so full, that that district produces
the greatest variety which is most examined."

I have hereafter noticed the snakes and the principal
reptiles here. Thousands of small frogs inhabit the
swamps, and afford an unlimited supply of food to the
different aquatic birds. We had three or four different
species; none, however, large. The commonest of all
was a very little frog, bright green, which used to sit
upon the caudock leaves and rushes, uttering a most
melancholy croak. But the deep regular clock of the
bull-frog, as we used to call it—which, by the way, is
a very small fellow for the noise he makes—is the
deepest and loudest of all. The frogs appear to come
into the swamps as soon as they fill; and I recollect one
year, when the swamps filled early, they first croaked
about April, which is the autumn here; and at the same
time I observed the fry of some small fish in one of the
lagoons. We had a curious-looking tree-frog, light yellow
brown, with very long legs, which lived in the bark of the

dead trees: the note resembled the setting of a saw, and we called it the carpenter-frog. I never recollect seeing a toad out here; nor did I ever notice a water-newt in any of our swamps or water-holes. Leeches abound in the streams and swamps: we used to catch them by throwing a sheep-skin into the water, and upon taking it up it was covered. We could sell them for a shilling a dozen.

I am nothing of an entomologist; but I was surprised that we saw so very few pretty butterflies out here. I only knew three or four species, and these were nothing extraordinary. The prettiest variety was one white, red, and yellow, which flew about the gum-trees. At times, a great cloud of moths, as large as small birds, would invade our tent in the evening, attracted by the light; they were remarkable, however, for nothing but their size. In the early summer, swarms of locusts, resembling a handsome beetle more than the locust of Egypt, settled on the gum-trees; and the whole forest would ring with their loud monotonous drone. Grasshoppers of different species cover the ground in the dry summer weather; and thousands of mole-crickets live in holes on the plains. The most curious insect here is the praying mantis, a species of grasshopper, with wings like leaves, about six inches long. There is a very handsome species of wasp, which used to come into our forests in the early spring, and burrow into the sand. I do not believe there is any native honey-bee in Australia; but swarms of the common domestic bees yearly leave their hives and fly into the forests, and there is now plenty of honey to be found in the old hollow gum trees. These forests abound in beetles, of various species and colours; and I have no doubt a man might soon get a fine collection of the *coleoptera*, by poking about the old rotten trees and dead wood which strew the ground.

I have often regretted that I knew nothing of botany. Although the wild-flowers here are not so large and gaudy as we generally see them in a southern land, it is when collected in a nosegay that their beauties strike the eye; and it is only then that we can form any idea of the delicate and varied tints of the little wild flowers which we pass by unheeded when growing on the plains and in the forests here. Some of the heaths and grasses are very fine; but there is a great absence of large wild-flowers in Australia.

Of course, such a country as Australia must present a wide field to the naturalist, let his particular taste be what it may; and the further we go back from the peopled districts, the more rare and, as yet, undiscovered species, especially of plants and insects, will be brought to light. These two branches of the natural history of such a country as this, must be only in their infancy; and it was always a matter of surprise to me that so much is already known of the general *Fauna* of this land; and I cannot close my slight sketch of the ornithology of this country, without paying a compliment to the perseverance and research of Mr. Gould, whose splendid work, which is, unfortunately, beyond

the reach of the field naturalist, is a pretty good proof
of the attention he must have bestowed upon the subject.   During the last three years, I believe only about
two birds unnoticed by him have been discovered.   As
to myself, I was never lucky enough to fall in with a
single new bird; and often have I been disappointed, in
my early days of collecting here, when I have taken up
a specimen which was new to me, and showed it to an
old collector, by his quiet remark, " Oh, yes; I know this
bird well : it is very common in such a district."

The collector here has many difficulties to contend
with: he will be able to procure very little assistance;
must depend almost entirely upon his own exertions;
and unless he has some small capital, will not be able to
give his sole attention to a pursuit which yields but little
present emolument.   I had to procure my daily bread by
my gun; collecting was with me only a secondary con-
sideration; and I was always obliged to be camped in
the settled districts, within reach of a market for my
game.

Collecting is not a profitable occupation; and this
is hardly yet a country where men care to give up
their time solely for the sake of benefiting science.   But
there are now many able and zealous naturalists at work,
and Melbourne can boast of a museum, raised within the
last three years, that is a credit to the curators.   Very
few bushmen or settlers care anything about the natural
history of the country.   A mob of cattle or a flock of
sheep are naturally of more interest to the squatter or

stock-rider than any rare animal or bird that he may
chance to fall in with during his daily rides; and the
only specimen that possesses much value in the eyes of
most of the colonists at the present day, is a lump of
quartz well inlaid with gold.

List of the ANIMALS, BIRDS, REPTILES, *and* FISH,
noticed in these pages.

ANIMALS.

Kangaroo.	Native Cat.
Wallaby.	Flying Mouse.
Wild Dog.	Kangaroo Rat.
Native Bear.	Bandicote.
Wombat.	Bush Rat.
Opossum.	Beaver Rat.
Ring-tail.	Platypus.
Large Flying Squirrel.	Spiny Ant-eater.
Little Sugar Squirrel.	Bats (3 species).
Tiger-Cat.	Fur Seal.

BIRDS.

Eagle Hawk.	Marsh Harrier.
White Fishing Hawk.	Swamp Hawk.
Peregrine.	Kite.
Hobby.	Carrion Hawk.
Merlin.	Large Owl.
Blue Falcon.	White Owl.
Kestrel.	Yellow Owl.
White Kestrel.	Grey Owl.
White Goshawk.	Brown Owl.
Sparrow Hawk.	Little Brown Owl.
Musquito Hawk.	Morepoke.
Harrier.	Little Morepoke.

Night-jar.
Laughing Jackass.
Sacred Kingfisher.
Kingfisher.
Magpie.
Black-backed Magpie.
Blue Jay (2 species).
Swamp Magpie.
Carrion Crow (2 species).
Blue-eyed Crow.
Mocking-bird.
Large Grey Cuckoo.
Common Cuckoo.
Bronze Cuckoo.
Summer-bird.
Miner.
Green Thrush.
Grey Thrush.
Mountain Thrush.
Wattle-bird.
Redwing.
Leatherhead.
Wood Swallow (2 species).
Honey-bird (3 species).
Bell-bird.
Spine-tailed Swift.
Common Swift.
Swallow.
Marten.
Mounting Lark.
Small Lark.
Meadow Pepit.
Rock Pepit.
Robin (5 species).
Swallow Diceum.
Blue Wren.
Emu Wren.

Brown Wren (2 species).
Sedge Warbler (2 species)
Satin-bird.
Stockwhip-bird.
Fantail.
Fly-catcher (2 species).
Great Shrike.
Stringy-bark Shrike.
Thickhead (2 species).
Oceole (2 species).
Crested Shrike.
Cobbler's-awl Bird.
Coachwhip-bird (2 species).
Wax-billed Finch (4 species).
Diamond Sparrow.
Diamond-bird.
Black Cockatoo (2 species).
White Cockatoo.
Yan Kate.
Corella.
King Parrot.
Lowry (2 species).
Rosella.
Blue Mountaineer.
Green Leek.
Swamp Parrot.
Ground Paroqueet (2 species).
Red-shouldered Paroqueet.
Swift-flying Paroqueet.
Common Green Paroqueet
    (2 species).
Zebra Paroqueet.

———

GAME BIRDS.

Emu.
Wild Turkey.
Lowan.

Native Pheasant.

Black Swan.

Magpie Goose.

Cape Barron Goose.

Mountain Duck.

Black Duck.

Wood Duck.

Pochard.

Whistle-wing.

Shoveller.

Teal.

Musk Duck.

Coot.

Dabchick.

Water-hen.

Bittern (2 species).

Heron.

Purple Heron.

White Crane (2 species).

Little Egret.

Spoonbill.

Straw-necked Ibis.

Nankeen Crane.

Spotted Bittern.

Native Companion.

Bronze-wing Pigeon.

Scrub Pigeon.

Ground Dove.

Crowned Pigeon.

Snipe.

Landrail.

Painted Snipe.

Common Quail (2 species).

Scrub Quail.

Painted Quail.

Nuthatch Quail.

King Quail.

Silver Quail.

Spur-wing Plover.

Plover of the Plains.

Golden Plover.

Stone Curlew.

Spotted Crake.

Lesser Crake.

Dunlin.

Black and White Stilt.

Small Stint.

Sea Snipe.

Large Sandpipers.

Bull-eyed Plover.

Curlew.

Avocet (2 species).

Crested Grebe.

Great black-backed Gull.

Lesser black-backed Gull.

Common Gull.

Caspian Tern.

Common Tern.

Black Tern.

Oyster-catcher.

Gannet.

Cormorant.

Diver.

Pelican.

Penguin.

Mutton-bird.

Petrel.

Little Stormy Petrel.

REPTILES AND INSECTS.

Black Snake.	Tarantula.
Carpet Snake.	Leech.
Whip Snake.	Blowflies.
Deaf Adder.	Ants.
Guana.	Musquito.
Sleeping Lizard.	Sand-fly.
Bloodsucker.	Locust.
Common Lizard.	Mole-cricket.
Bull-frog.	Wasp.
Green Frog.	Butterflies.
Carpenter-frog.	Moths.
Centipede.	Praying Mantis.
Scorpion.	Grasshoppers.

FISHES.

Eel.	Herring.
Brim.	Mullet.
Black-fish.	Murray Cod.
Trout.	Fresh-water Cray-fish.

SEA FISH.

Snapper.	Fiddler.
Flat-head.	Flounders.
Sea Pike.	Butter-fish.
Garfish.	Jelly-fish.
Salmon.	Cuttle-fish.
Salmon Trout.	Star-fish.
Cray-fish.	Crabs.
Oyster.	Toad-fish.
Shrimp.	Porcupine-fish.
Dog-fish.	Mutton-fish.
Shark.	Limpets.
Benito.	Saw-fish.
Stingoree.	Shells.

CHAPTER X.

NOTICES ON SHOOTING—SUGGESTIONS FOR THE PRESERVATION OF
THE GAME—THE DOGS OF THE NATIVES AND BUSH DOGS.

NOWHERE do we see better guns out in the field than
here, and a man is not known among the shooters by his
personal description, but by his gun: "Used to shoot
with an old double Joe Manton;" "Had a long single
Rigby;" or, "Shot with my Purday gun;" was the
manner of speech by which a stranger was recognized
among our mob. Any one who keeps his eyes open has
every chance of picking up a good gun here cheap, for
nearly every emigrant brings one out with him, often a
first-class gun, perhaps an old family relic, or the parting-
gift of some old sporting friend; and this is generally
the first thing that goes when he becomes hard up on
landing. I suppose it is owing to a different class of
men coming out, but I have remarked that we do not
see so many first-rate guns brought into the colony now
as formerly.

There are several good gunmakers in Melbourne, but
all colonial work, especially little jobs, are dear.
The best powder costs 5s. per lb.; shot 6d., caps 7s.
per 1,000. Some wretched rubbish is sent out here
in the shape of powder, and if the shooter happens

to run out in the bush, he will most likely have to put up with the common treble F., at 5s. per lb. There is no saving in shooting with cheap powder, for, independent of the foul state in which it keeps your gun, and the wounded birds that go away, a pound of Hall's best No. 2 grain will go as far again as the weaker powder. A man who shoots for his living cannot be too particular in the choice of his ammunition, for in duck-shooting he can easily lose more by wounded birds in the day than will keep him in ammunition for a week, and no one but he who has experienced them can judge of the duck-shooter's feelings when his cap misses fire, after having crept on his hands and knees up to a flock of black duck, for perhaps half a mile, through a wet swamp. Pigou and Wilks's No. 2 grain was my favourite powder whenever I could get it; but all the good brands are pretty much alike, if you are only certain that it is the genuine article, and the canister fresh. The best plan is to stick to one gunmaker, and he will generally use you well.

It cannot be denied that the game is rapidly disappearing in all the settled districts, especially near town, and if steps are not speedily taken to prevent the wholesale destruction of the birds in the breeding season which is now carried on, in a few years the shooter's occupation in Victoria will be gone. Much as we may all object to the principle and working of the British game-laws, it is quite certain that, until the law interferes and makes it an act illegal for all, there can be no preventing it. For, however well one man may be disposed, and wish to shoot fairly, it is hardly likely that he will care to spare the breeding-birds, when he knows that they are pretty certain to be shot by some one or other less scrupulous than himself.

The inhabitants of any wild country, who depend upon the chase as a subsistence, have, as it were, a *primâ facie* right to the game of that country, and are, perhaps, justified in taking it at any season, as they best can. They wander about from spot to spot, and are not continually disturbing one district; they have different hunting-grounds for different seasons; their implements of chase are rude, in comparison with those used by the civilized man; and they never care to take more than just enough to satisfy their wants, and there would be little fear of the game being ever entirely killed out if they were the only persons who followed the chase. But the case is far different when thousands of strangers flock to a new country, and wage an indiscriminate war at all seasons against the wild game peculiar to that land. It is then time that some measures should be taken to preserve the game, and if the shooters themselves are too blind to their own interests to do so, the law should interfere. But let me not be misunderstood. I am not here advocating any system of game-laws that will cramp the sportsman in this free country as the Legislature has already done at home; all I wish to see is a stop put to the ruthless slaughter of the old birds in the breeding season, and I am sure every fair sportsman will join in my views. Let us have no license. Let a man still be free to wander where he will on land that is not purchased, but let us have proper seasons fixed for killing the game. What is sauce for the goose would then be sauce for the gander—it would be as fair for one as another—and all who take a pleasure or feel an interest in field-sports out here would be equally bene- fited. When the game of any country becomes a marketable article, and of sufficient value to induce men to devote their whole time to its pursuit as a means of gaining a livelihood, it should in some measure be pro- tected, especially as it is not private property. One might imagine that it would at least be the interest of the shooters themselves to do so—at all events to spare the goose that lays the golden egg; but, un- fortunately, what is everybody's business is nobody's; and, although they are the first to complain when they find the game decreasing, not one will give himself the slightest trouble to keep up the breed. It matters little to me—I never expect to have another head of game out of Victoria—but I have seen enough in five years' shooting to prove that, unless some steps are speedily taken to preserve the breeding-birds, in a few years none will be left to protect.

The game list of Victoria should include those birds that are bought up as game, such as the turkey, the ducks, the pigeon, the quail, the snipe, and the rail. There are several other species quite as good for the table as these, but which are hardly considered game, viz., the bittern, the coot, the heron, nankeen crane, plovers, and wattle-birds. It is a matter of doubt whether the kangaroo will ever be deemed worth preserving, but I have already touched upon this subject.

Some of the game birds, such as the quail, pigeon, and rail, only come into this part of Port Phillip to breed, and I know the shooters here will say, that as soon as they have done breeding they all leave with their young, and unless shot just at the times when they come down none would be got at all. This may be partly true, but, in my opinion, we never gave the birds a fair chance to see how long they would stay on the breeding-grounds with their families after the breeding season. I fancy, if allowed to breed in peace, they would remain till the end of autumn, perhaps well into winter. As to snipe, they might be killed at any time when they can be found here, for I fancy they breed up in the ranges early, in places little trod by the foot of the white man, and those which do come down into the peopled districts are the old birds and birds of the year. The ducks pair off to breed about the end of September (I once took a nest as early as August), and the flappers come down to the creeks in January. The pigeons breed in December, and the young birds are flyers by the end of January. The heart of the quail breeding season is early in December, and in January we kill strong flyers. My opinion is, therefore, that the safest way to preserve the game here, would be to make November, December, and January "fence months," for every species of game excepting snipe, in the settled districts. The shooters would then have a little good quail-shooting when they first came, and before they settled down to their breeding haunts. February and March would be the best months for general shooting; duck and kangaroo would keep them employed during the winter, and it is, indeed, hard if they could not afford to give the birds a three months' rest out of twelve, especially as they would reap the benefit of it themselves, and as during the hot summer season shooting is far more a toil than a pleasure, and half the game is spoilt by the heat.

I am not here alluding to the professional shooter alone. It would be far more satisfactory to those sportsmen who merely follow the chase as an amusement, for they might then be always certain of a good day's sport within an easy distance from town, which is very doubtful now, when men are roaming over the country with guns, disturbing the birds during the whole of the breeding season.

I suppose there is already some law of trespass out here, but I don't know how it stands. This, however, I do know, that there is always bother enough if by chance the shooter enters a private paddock, especially near town. For my part, I hated the very sight of a three-rail fence; and half the pleasure of shooting in this wild country was taken away whenever I had to enter an enclosure, or ask leave to beat for my game.

It is very properly prohibited to shoot the ducks which resort to the Botanical Gardens in Melbourne within a certain distance of the enclosure, in fact, to shoot at all within the boundaries of the town; and there is a fine for shooting on a Sunday, which is strictly enforced in the neighbourhood of town, but has hardly found its way yet far into the bush.

I should much like to know whether the aborigines of this country originally possessed any particular breed of dog for the chase before the stranger landed on these shores, or whether the domestic dog was introduced into this country by the white man. Unlike other savages, I do not believe that the Australian natives depended at all upon the dog for their success in the chase. Their original methods of hunting prove this; for although they certainly now do prefer a gun to a spear when they can get it, and their dogs assist them much in killing kangaroo and opossum, they still stick to their primitive habits of the chase, especially in the wild districts. For instance, they will encircle a mob of kangaroo, and kill them with a spear or a waddy; they will stalk the wild turkeys on the plains under cover of a bush, which they carry before them, and snare them with a noose on the end of a long pole; they will watch a creek for hours, hidden in the rushes, and when a mob of ducks pass by, will knock down two or three with a waddy or bomerang; and they can also spear the ducks; they will sit by a water-hole on a summer evening as motionless as statues, and snare the pigeons that come down to drink. They have peculiar methods of catching quail. They can tell, by examining a tree, whether an opossum is at home; and they soon run up, by cutting

nicks in the bark, as a purchase for their fingers and
toes, with a tomahawk (which, before the white man
settled here, was made of stone, and answered every
purpose), and drag him from his hole. Of course, no
white man can ever equal them in stealthily creeping
on to their game; and I have often remarked that
neither the kangaroo in the forest nor the wild turkey
on the plains take half the notice of a black that they
do of a white man. As all the species of birds and
animals above mentioned, and fish, formed their principal subsistence in the way of the chase, they could get
along very well without dogs; and as they had nothing
to fear from the attacks of any wild animal, and no
property to protect, I think it most probable that the
domestic dog was introduced into this country by the
white man, and that before he landed the blacks did
without their assistance; for I cannot believe that the
wild dog could ever have been broken from a state of
nature to become of any service to man, more than the wolf
of Europe. Still the oldest settlers seem to have no recollection of seeing the blacks without dogs. That they are
very fond of them, is evident from the pack which accom-
panies every tribe — hungry, mangy, sneaking-looking
curs, of no particular breed; most likely a cross of every
blood known in the colony. I have seen a Lubra, or
native woman, suckling two puppies; and, like monkeys,
these ladies have a particular fancy for fleaing their dogs.

Next in relation to the bushman's mate stands his
dog; and I should almost feel myself wanting in grati-

tude were I to pass over these faithful companions without a slight notice.

It is difficult to say what is the most general breed of dog we meet with in the bush : in fact, we rarely see a true-bred one at all.   Every bushman probably brings a dog or two into the bush with him, of such breed as he fancies best; and as there is no restraint and no care bestowed in crossing them, they breed indiscriminately, and it would puzzle a good dog-fancier to distinguish one breed from another.   But mongrels as they are, these bush dogs are not to be despised.   Although self-taught, nature supplies the place of education;   and their natural instinct seems to be much more highly developed than among the fuller broken and truer bred dogs of the old country.   Every bush dog is a sporting dog after his own fashion ; and as there was no tax, and their keep in the forest cost nothing, they must have been the veriest curs that were turned out of our kennels.   Our dogs were used for every purpose;   and as they were treated more like companions than servants, they appeared to identify themselves with us in every transaction, and seemed to fancy a day's shooting was got up as much for their pleasure as our own.   It is little wonder that they were keen after kangaroo and opossum: we never gave them any meat except what they helped to kill themselves, nor did they seem to expect it.   They knew where the offal of the kangaroo we had killed in the day lay in the forest, and regularly every evening went off to feed; and if there was no

kangaroo handy, they would stroll away from the tent at
night, run an old " 'possum up a tree," and stand barking
under it till they brought one of us out to shoot it.   No
kennels for them; each had its own little den under
some old tree-root close to the tent, and we always
slept in perfect security.

It is dangerous to go up to a bush tent or hut after
dark, on account of the dogs; and the best plan, as soon
as one sees the camp-fire, is to " Coo e, e," to warn the
inmates of one's approach.

I can often fancy a shooter at home seeing the turn-
out for a day's sport here.   In my shirt-sleeves, with a
game-bag on my back, and my pack of mongrels at my
heels (for no matter, whatever was the sport all the dogs
were sure to follow us), one a half-bred bull and terrier,
a large half-bred mastiff and hound, a fine-bred grey-
hound terrier, and a long-backed spaniel, worth, in the
eyes of an English sportsman, to use an old phrase of
the road, " about ninepence a side, pick 'em all the way
round," I looked far more like a rat-catcher than anything else.   Yet these were the dogs upon whom we
depended, not only for our personal safety, but our daily
bread.   Little fear of any one molesting us at night with
these protectors round us; and as for sporting, they were
all close-hunting dogs for quail and snipe—would retrieve
a black duck from the thickest rushes; and, in " fur,"
scarcely anything, from a kangaroo or opossum down to
a bandicote or bush-rat, escaped them.

There is a small tax now of 1s. 6d. per year laid upon

every dog kept in and about Melbourne — a kind of *douceur*, I fancy, to the police; and each dog must be registered and wear a collar. Although fewer diseases prevail among the dogs out here than at home, I have heard it remarked, that they are far more difficult to cure. Hydrophobia is unknown. The worst and most common sickness is a species of distemper, not at all like the distemper at home, but a kind of spasmodic affection in the loins, which comes on at all ages. It is to be cured; but I never knew a dog worth much after it. The receipt which I got from a sporting " Vet " was to cut the roof of the mouth across, from gum to gum, so that it may bleed freely, and give a dose of garlic every morning for four days. I do not know what it is owing to, but dogs either get very soon worn out in this country, or very cunning; for a sporting dog is worth little or nothing after about his fourth season.

How imperceptibly and closely does a man become attached to old localities, and old companions, even if they are but dumb animals; for, childish as it may appear, it was with feelings of deeper emotion than any one can imagine who has not, like myself, spent year after year in the solitude of the bush, that I parted from my mates and the old bush-tent, and for the first time in my life drove my dogs back, who followed me when I left with half-imploring, half-incredulous looks. As I turned my back upon the forest, I felt that my sporting career in Australia had ended; that I was parting with friends whom in all probability I should never see again; that I was leaving a home at least, even if it was a humble one, with no consolation in the reflection that I had to seek another home and new friends in whatever strange land the wanderer's lot might cast me.

CHAPTER XI.

THE SNAKES AND REPTILES PECULIAR TO PORT PHILLIP, AND OTHER BUSH ANNOYANCES.

ALTHOUGH the bushman has nothing to fear out here from the attacks of any wild animals, he has still his secret enemies, which in many cases are as dangerous as the open foe; and what he has most to dread in the Australian bush are the snakes. I do not believe any part of the world can be more infested with these reptiles in the summer season. Let him walk where he will—in the depths of the forest, in the thick heather, on the open swamps and plains, by the edges of creeks or water-holes—the shooter is sure to meet with his enemy, the black-snake. It enters his very tent or hut, and coils itself in his blankets. In fact, nowhere is he safe; and if he did not altogether banish the thought of them from his mind, he could never have a moment's peace. It does, indeed, appear as if the eye of a watchful Providence peculiarly guarded the traveller in these wilds; for at any moment he is liable to tread upon a deadly snake, coiled up in his very path, which does not always get out of the way, but lies watching him with his basilisk eye, ready in a moment to make the fatal spring if touched, and very often the snake is not seen till the

danger is past. Much as I was accustomed to the sight
of them, and the hundreds that I have killed, I never
saw one without a cold chill running through my blood;
and it is often with a shudder that I look back upon the
many narrow escapes I have had from snakes. How
I avoided being bitten is a mystery to me. I once
threw myself on my blankets for a rest, during a hot
windy day, in my shirt-sleeves, and a large carpet-snake
lay curled up within three inches of me! Twice have I
taken up the little whip-snake in a bundle of dry grass;
and twice have I had a large snake twist itself round my
leg; and in one instance my leggings saved me, for the
snake struck me below the knee. I have picked up
a dead quail in long grass, which had fallen close to
a snake; and scores of times have I all but trod on them
in thick grass. I always wore long boots, or gamekeeper's leather leggings in the bush during the
summer; and I should recommend every one to do
the same. I consider the greatest danger we ran was
if we chanced to pick up an old log at night for the
bush-fire in which might be a snake; a man cannot be
too careful in handling dead logs and sticks in the
forest; for, independent of snakes, this dead-wood is
infested with centipedes and other insects, the bite of
which is dangerous. One thing is fortunate, by constant
practice the eye becomes so accustomed to range over
the ground that, in most instances, I could see a snake
before I reached it, unless it was coiled up very snugly.

I could never identify more than three distinct species

of snake out here : the black-snake, the diamond or carpet snake, and the little whip-snake; all, I believe, equally poisonous in their bite. We had many varieties in colour, but I think these were the only three distinct species.

A small kind of boa is met with up the country, according to the Blacks, which is harmless; but I never saw it.

There is another species in some parts of the country which they call the deaf or death adder, but it is never met with in the districts where I have been. It is described as a short, blunt snake, with legs, and as being able to sting or bite at each end : and it is said to be the most deadly of all the snakes. That such a reptile as this exists, I never can believe; although there is a species of short, thick snake, unknown in these parts, found on some of the dry stony rises, and this is probably the one meant. I have heard the most extraordinary stories respecting the size and quantities of snakes met with in some parts of the colony; and this, and the wonderful escapes they have had, is a prolific subject with some old bushmen, but I always received such "yarns" with the greatest of caution.

The black-snake is the handsomest, but certainly the most venomous and spiteful, in appearance of the whole lot. It is of a rich black colour, above the belly-plates light. We had a variety in which the belly-plates were copper-red, which we called the copper-snake. It was always smaller and thinner than the common black-snake, and might have been the young. None of the snakes here run to a very large size:—five feet will perhaps be about the average length. The largest I ever saw was a black-snake, killed by my mate in a thick scrub on the beach, near Mordialloc, which we called the two-mile scrub, certainly the worst place for snakes that I knew. It was six feet five inches long, and very thick. On showing it to a Black, he observed, "Ah! me know that fellow long time." I think both the black and the carpet snake were equally common with us. We generally used to find the black snake more among the timber and thick scrub than the other; but in the dry season we were sure to find both near water. There is a strong scent peculiar to the Australian snake, and I have often smelt one long before I saw it.

The carpet-snake runs much about the same length as the black-snake, but is rather thinner. I generally found them in more open places; and often on the plains in dry weather, they would lie coiled up in a crab-hole, or print of a bullock's hoof. The carpet-snake is of a brown colour, with a yellowish tinge and light belly, the shades varying much, according to age and season. It is a dangerous plan to let heaps of glass bottles accumulate near a bush-tent, for they attract snakes much in hot weather.

The little whip-snake is the smallest of all, being hardly thicker than one's finger, and rarely over a foot in length. It rather resembles the blind-worm at home in colour and appearance, but it is longer, and the tail more pointed. They frequented the dry plains, were local, and I often used to find them under heaps of dry cow-dung.

The snakes here lay up during the winter in old logs, dead-log fences, and holes in the earth. They disappear about the end of March, and come out again in September. They say here, that in the end of February is the pairing season, and then they travel by night. I cannot say if this is correct, as I never killed one at night, except in a log. At all other times they retire as soon as the sun goes down. They are the most dangerous when they first come out, for then they lie in a half-torpid state, and don't care to get out of the way. One thing is certain, that the snake will rarely if ever molest a man, unless trod upon, or so hard pressed that it cannot get away. They generally glide off out of sight, or if they do lie still it is in hopes of not being perceived. They can hear the approach of a footstep a long way off; it is wonderful how quickly they disappear. I have seen a snake lying in a bush, and have only taken my eye off it for an instant, to see if my cap was all right before I fired, and it has vanished as by magic. I have seen some persons kill them with a stick; I always fancied a charge of shot was the safest, and I rarely went out in any day in summer without killing two or three. It is best to approach a snake sideways, for they say here that they can cast themselves backwards as well as forwards. I never but once saw one make a spring, and that was at a dog. The snake was in a half-erect position, and shot out its full length like lightning. Many dogs are very quick at killing snakes, and will seize and throw them up like a rat, but sooner or later they pay the penalty of their rashness. An old bush-dog generally stands over a snake, at a respectful distance, and barks till the shooter comes up.

The laughing-jackass and stump-lizard both destroy snakes; and they say that Underwood, in Van Diemen's Land, who cured the bite of snakes, discovered the secret of his elixir by watching a battle between a snake and a stump-lizard. After the lizard had killed the snake, he saw it eat the leaves of a small plant. He gathered some, made a decoction of it, and this was the secret of Underwood's mixture. Whether this was the case or not I am unable to say, but I believe his remedy is very efficacious; and he has himself acknowledged that the principal ingredient is a plant on which we tread in this country every day of our lives.

It is singular, considering how much I was always about the bush, and the number of snakes that I killed, not a single instance of a snake-bite ever came under my actual observation; so it appears that these snakes are less to be feared than might at first be imagined. I have known men who have been bitten and recovered, so that the bite is not always fatal. Much, I think, depends upon the state of the blood, and the season of the year. One man I knew was bitten in the finger by a whip-snake, when putting up a fence. He coolly laid his finger on a post, and chopped it off with his axe, and thus probably saved his life. My remedy, if I had been bitten, would have been to cut the wound till it bled well, and put on it a charge of powder, and flash it off. I think this might have stayed the poison until I could have reached medical assistance. Many carry a piece of caustic; but the new remedy, I believe from India, where it has been tried and found most efficacious, is ipecacuanha. If bitten, score the wound with a penknife till it bleeds, make a paste with a little ipecacuanha and spittle, and bind it round the wound. Of course these are only temporary remedies till medical assistance can be obtained; but when we consider how liable any one is to be bitten out here, it certainly would be prudent for every one to carry a little ipecacuanha in his pocket, even if he never wanted it. The Blacks have a remedy, and no doubt it is herbal.

The snakes here live always on and in the ground, and not in trees, which, however, they can climb, for they are not unfrequently found in a magpie's nest. I was once standing quietly by a creek, watching for ducks, on a summer's evening, when I heard a rustling in the scrub, on the other side, and I saw a large carpet-snake, swarming up a tea-tree pole, and presently another and another, till I am certain there were at least a dozen crawling up and down the poles at various heights. I did not stop to see what they were about; as the Yankees say, I soon " made tracks back," for I fancied I must have come upon a snake settlement. I believe there are such places, where hundreds congregate in long grass or thick scrub. I think a great many of the bullocks that lie dead along the plains in the summer are killed by the snakes. The sheep often kill them by jumping with all four feet upon the snake.

The Blacks are very timid about snakes; yet, although they travel bare-legged and bare-footed through the bush at all seasons, they never tread on one; in fact, their eyes are like an eagle's, and they can see anything on the ground in an instant. They are very careful, however, in getting over a log, rarely treading on it. They will eat snakes, which they kill themselves, when they are certain the snake has not bitten itself, which it often does in its dying agonies. I have eaten the black-snake, and had it been a little fatter should not have known it from eel.

The principal food of the snake is small animals, birds'-eggs, and frogs. I once saw a large carpet-snake charming a lot of birds. It was under an old honey-suckle, which had been blown down, and a congregation of small bush-birds were gathered round it, hopping, chattering, and fluttering about the dead branches of the tree. The motions of the snake were the most graceful I ever saw: it was half-erect, moving its head backwards and forwards, shooting out its tongue, evidently endeavouring to decoy a victim within reach, which it would soon have done, but it caught sight of me, and glided away, and the performance stopped.

There is no real water-snake in this country, but all the snakes can swim, and in summer are always on the edges of wet swamps, creeks, &c. I have often seen a snake drinking, when I have been watching by the side of a water hole. I once shot a pair of ducks in a creek, and they fell in the rushes on the opposite side. As I had no retriever, I stripped and swam in; and while I was swimming across, I saw what I took to be a black piece of stick, lying on the top of the water; when I came up to it it proved to be a large black-snake, lying, perfectly motionless, at full length on the water. I passed within a foot of it, but it never moved. I often wondered since whether it was after the ducks.

The bush-fires must destroy thousands of snakes annually, and wherever the country becomes cleared they will, of course, in a great measure disappear. But they can never entirely be rooted out of this land, where so many miles of swamp, scrub, and heather must for ever remain, in their original wild state, a harbour for the snake and other vermin. But I think Government should offer some slight reward, say sixpence or a shilling, for every snake's head that was brought in.

It is strange that this country should be so prolific in snakes, while in New Zealand, only about a thousand miles distant, not a reptile is, I believe, to be found.

The guano is a large species of tree-lizard, and frequents gullies and ranges where the timber is high, and the localities wild and unfrequented. It was very rare in our district; in fact, it is found only in the most solitary places. The guano runs from all sizes, up to ten feet, and I have heard of them even longer. The body is thick, covered with a close scaly hide, of a dark brown colour; the head is large, and the tail long and thin, like a whip-thong. It is a repulsive-looking reptile, and I must say I never liked the sight of them. A friend of mine once met a large guano, in a narrow bush-track, marching along with a great piece of beef in its mouth, which it had stolen from a tent. It carried the beef with its head in the air, like a retriever carrying a pheasant. It dropped the beef when he fired, and disappeared into the bush; so, after all, he got the best of the bargain. The guano is not venomous, but can bite severely, as the scars on the faces of those dogs that hunt them will testify. It is very nimble, and can run up a tree like a cat, keeping its body out of sight behind the trunk, or a large limb, peering down, with its hideous countenance, on the shooter below. It is almost impossible to shoot them unless there are two guns, one on each side of the tree. The guano is eatable, and the tail a bush delicacy. I have seen them in the Dandenong ranges, and I believe they are very common in the high timber on the Gipps Land Road.

The *Sleeping* or *Stump-lizard* is another repulsive-looking but inoffensive reptile. It runs to about one foot in length, is very thick in proportion, and the tail is short and blunt. It is of a variegated brown colour, the belly livid blue, the inside of the mouth and tongue black, and the belly is not covered with plates, as in the European lizard, but the skin is continuous. They are common all over the bush during the summer, frequenting generally moist situations,—such as tea-trees and damp grass; sometimes, however, in dry heather. The sleeping lizard well deserves its name. It is the very counterpart of its relative, the guano, being sluggish and lazy, always lying apparently half asleep, did not the bright little eye prove that they are " wide awake." They never try to escape by flight when a dog attacks them; all they do is to turn their head towards it with open mouth, and I fancy their repulsive look often protects them. Dogs will set them like quail. Their principal food appears to be grubs and caterpillars; and as these insects require very little catching, I should say the life of the stump-lizard is about as lazy a one as any in the colony. They can bite severely for their size.

There are several other species of small lizard in the bush, all harmless; but one they call the bloodsucker—a perfect guano in miniature—they say, is poisonous. I fancy not. But a stranger in a foreign land should always be careful in handling reptiles, unless he well knows their habits.

It was curious that we had no alligators in this part of the country. Many of the creeks would be the very places for them, and I am sure some of the swamps are wild and dismal enough to hold any kind of uncouth reptile. I believe the Blacks fancy that some species of large reptiles or amphibious animal do inhabit the large swamps, and I have often had the same opinion myself, when camped on the edge of one of these dreary, im-penetrable marshes. I have listened to the extraordinary noises that issue from the reeds and scrubs at night. Surely, I have thought, there must be some reptiles in these wilds unknown to us, and perhaps, after all, the bunyip may be no fable.

Besides the snakes, we had many other little annoyances during the summer months in the bush, which it was not always easy to avoid. Centipedes, six inches long, are to be found in every old log, and under stones. Small scorpions abound on every stony rise; and tarantulas, or large spiders (as the bushmen call them, triantulopes), as big as penny pieces, with a dozen great hairy legs, come crawling down the sides of the tent in wet weather; and the bushman should always well examine his blankets before he turns in. The cruellest practical joke I ever heard of was played in a bush-tent, when one of the party laid a dead black-snake in the bed of his mate. The man upon whom the trick was played did not die from the fright, but his intellects received such a shock, that in all probability he would remain a lunatic for the rest of his life. The swamps are full of leeches, and very pleasant it is if one finds its way into your shoe when up to your knees in a swamp, crawling on to a mob of black-duck. The forests literally swarmed with great blue and yellow blowflies, which shoot out living maggots, and " Catch-'em-alive-oh" would drive a roaring trade among the clouds of little black flies that infest the bush during the summer. Ants of every variety and size, from the little sugar-ant up to the great red soldier-ant, ply their busy trade in the summer all over the forests. There is no keeping these little busybodies out of the tent, and it is no joke if a great bull-dog, or soldier-ant, about an inch long, finds its way up the leg of your trousers. Swarms of mosquitos hover over every marsh and water-hole in the evening, and myriads of sand-flies, hardly perceptible to the naked eye, are sure to attack the back of your neck and ears when seated by a water-hole, quietly watching for a duck or pigeon at evening. These are the only insects I really cared for. I could generally keep the mosquitos off by the smoke of my pipe, but with the sand-flies I could do nothing. The Australian bug is harmless, luckily, for it is about the size and shape of an almond; and as for fleas, they breed in the sand, so that it is easy to guess their name is legion. The blight in the eye, which is so common here during the summer, especially on the diggings, is brought on, I believe, by a small fly. The sting of the scorpion and centipede are not only very painful, but very danger-ous. A little sweet oil promptly applied soon cures the bites of the others.

CHAPTER XII.

BUSH LIFE.

MANY persons consider the shooter's life a lazy one, and are too apt to set down the whole of our " respectable corps" as a body of " loafers." Don't you believe it. No man can be idle who has to earn his bread first, and then cook it, before he eats it; and if any one doubts my word, he had better go and try it.

There are three or four classes of shooters out here. The " swell," who now and then comes out from town for a day's sport, and obtains the services of some professional shot, who knows the ground, to help to fill his bag. Money is no object with him, his sole aim being to take home a good lot of game, which he does not forget to show to his friends as his own killing.

There is a second class, who are very good at " shooting over the pitcher." These are, for the most part, old hands, men of sporting habits, who are generally to be found at the bar of a sporting public-house, where they " pitch," to any one who will stand nobblers, about the lots of game they used to kill, and wind up by abusing the new chum shooters, and the sporting prospects of the colony at the present time, as compared to their day. These men deal much in mysteries, and almost every one claims the honour of being the oldest duck-shooter in the colony, as I have heard at least a dozen diggers affirm that they sunk the first hole on Bendigo.

Then there are others, genuine sportsmen at heart, who know how to find game, and what's more, kill it when found; but being tied much by their business in town, have little time to devote to field sports, but who enjoy a day when they do have it, doubly, on account of its rarity, and it is a pleasure to go out with them; and besides them, there are others who, although settled, and following their regular trades, occasionally take a turn with the gun when business is slack, thus combining pleasure and profit. They never go out except when game is well in, and one night from home is about their limit; and these are the men who really enjoy the pleasure of sporting, without the hardships which fall to the lot of the regular shooters.

But the men whom I consider the real shooters are those who stick to it year after year, rough weather and smooth, no matter whether game is plentiful or scarce, trusting solely to their guns and their own exertions for a living; and, depend upon it, these must be anything but idle men.

It is astonishing how soon a man, who is made of the right stuff, can settle down to the rough usages of a bush life, and quite forget the domestic comforts he has left behind him in the Old World; and I have remarked that those men who grumble least—in fact, make the best bushmen—are often they who have moved in a better

sphere of life at home. With a good mate, as long as
his health stands, I do consider the shooter's life one of
the happiest and most independent in the colony. A
good waterproof tent properly put up, with a fly on the
roof, and a turf chimney, is by no means a bad residence,
and quite as warm and comfortable as half the weather-
boarded houses that are knocked up here. The shooter
is generally camped amidst the most beautiful scenery,
close to some good water-hole or creek, with plenty of
wood at hand. He has few artificial wants, and the real
necessities of life are easily and cheaply obtained. His
meat, of course, he procures by his own gun; and a bag
of flour, a little tea, sugar, and tobacco, fill his larder.
His cooking is simple, his furniture home-made. His
time is fully occupied, and not an hour hangs heavy on
his hands. His method of life is laid down by no rule.
He eats when he is hungry, sleeps when he is tired, and
works just when he pleases. The laughing-jackass calls
him up in the morning, and the flute-like note of the
magpie is his vesper bell. His very occupation preserves
his health. Content and health go hand in hand; and
although he has his share of the world's troubles—and
what class is exempt from them?—he has also the inward
satisfaction of feeling that he is leading a happy, independent life, and has no one to thank but himself for
his daily bread.

I have lived at times by myself in the bush, and it was
then a lonely, laborious life. Often have I toiled from
sunrise to sunset, come home dead beat to my lonely

tent, and, after ten hours' fasting, had to light my fire
and cook my solitary supper; and often have I turned
in fairly "baked," and put my supper off till morning.
But with a good mate, the case is different. It is true I
have spent many a rough day in the forest; and many a
night, when lost, I have lit my pipe, and thrown myself
down to sleep before a log fire: no companions but my
dogs, no covering but the sky, and no supper but an
opossum or bandicote thrown upon the ashes. And the
shooter should never leave his tent without a few matches
and a little salt, for he never knows where he may sleep
at night, or of what his supper may consist. But I can
also truly say that some of the happiest hours in a life
which certainly has had its bright as well as gloomy pas-
sages, have been passed in my bush-tent, when, after a
good day's sport, supper finished, and pipe lit, I have
thrown myself on my opossum rug, and the toils of the
day fairly over, have spent the hour before turning in
yarning with my mate over "the days past, the present,
and the future." At such a moment I would not exchange the rough freedom of the shooter's life for the
best situation in the colony.

The only thing he has really to fear is illness; but,
happily, few disorders prevail in this favoured land, and
these are chiefly confined to the towns and diggings, and
two-thirds of them the result of intemperance and a
reckless habit of life. Except in cases of accident, we
rarely hear of a bushman being laid up. Sickness will,
however, at times, enter the bush-tent, and on such occa-

sions a general gloom overspreads the whole of the little community. Far away from medical aid, the sufferer has to trust to such simple remedies as are at hand, and patiently await the issue. It is now that the rough sympathies of his mates are fairly awakened, and each one vies with the other in assisting and consoling the sick man. A hardy constitution generally " pulls him through ; " but when his complaint is beyond the help of man, he calmly resigns himself to his fate, and dies " unhonoured and " in many instances " unknown ; " for very often a man in this country knows very little more of his mate than his name. If visions of youth and home do flit across the mind of the dying man, it too often happens that there is not one among the strangers who stand around his death-bed with whom he can intrust a last message to his relatives and friends in the old country, who will probably wait month after month with " the sickening anxiety of hope deferred" for tidings of the absentee, which will now, perhaps, never reach them. This is one of the darkest pictures in bush life, but it is one which, in the early days of the diggings, was too true. I have seen death in more shapes than one in the bush, and it is then, and then only, that a true sense of the loneliness of his life breaks fully upon the wanderer's mind ; and as he misses his old comrade from the evening bivouac around the camp-fire, he smokes his pipe in silence, thoughts of his own happy home in early days will pass, like bright but transient gleams of sunshine across the field of memory, and for a while his reckless spirit is subdued by feelings of a deeper and more serious caste.

It is strange that the man who lives by his gun rarely saves any money; and this is the reason (whatever may be my own inclinations) why I should scarcely be justified in recommending any one to follow this life who comes out here with the hope—too often a delusive one— of making a rapid fortune. There is very little forethought with the shooter; and I suppose that the old law of the rolling-stone gathering no moss holds good in his case as well as any other. If he makes his daily bread he is content, for he seems to think with Burns,

> But cheerful still, we are as well as monarch in a
>   palace, O ;
> Though Fortune's frown still hunts us down, with all
>   her wonted malice, O ;
> We make, indeed, our daily bread, but ne'er can
>   make it farther, O ;
> But, as daily bread is all we need, we do not much
>   regard her, O.

Nowhere do we meet with more real friendship and genuine kindness of heart than in the bush. Rough in aspect, careless in dress, off-hand in his manners, there is a vein of simple and warm-hearted kindness running through the character of the real bushman, which we rarely, if ever, find among men whose better feelings have become insensibly deadened by a continual intercourse with the world. Isolated, as it were, from his fellow-men, solely dependent upon his own exertions for his daily bread, he feels himself under no obligation to any one, cares little to form new acquaintances, and always appears reserved and shy before strangers, especially "new chums;" but let him fall in with an old mate, or man of his own stamp, and the meeting is often of a boisterous character. No one more readily sympathizes with the reverses of a mate; and so little selfishness is there in his nature, that he willingly shares his all with him, whether it be his last shilling or his last fig of tobacco. His rude hospitality is proverbial; and the benighted traveller always finds food and shelter at the bushman's tent, as a matter of course; and, unlike the way of the world in general, the more "hard up" the stranger is, the more he is welcome. This is all done without ostentation, as a duty he owes to his fellow-man, and upon the principle that any day or night he may require the same assistance himself. I am here alluding to those men who knock about the bush on their own resources, living by wood-splitting, shooting, &c., and not to the regular settlers on stations; although, for my part, I can say that there were but few stations which I called at where I was not welcome to such accommodation as the "men's hut" afforded.

CHAPTER XIII.

GENERAL REMARKS ON THE SCENERY, CLIMATE, AND THE SEASONS
OF PORT PHILLIP.

THERE is a monotony in the scenery of this part of
Australia which is very wearying to the eye; and
although at times the traveller suddenly comes upon
a break in the landscape, the beauty of which no pen
or pencil can portray, yet the thick forests and the low
swamps and plains are of such vast extent, that the
wayfarer in Victoria may plod on for many a weary mile
with one unvarying landscape continually before his eyes.
Deep forests of gum and stringy-bark, evergreen both
in summer and winter; flats of stunted honeysuckle,
bearing no resemblance but in name to the sweet wood-
bine at home; parched and barren plains, miles in extent,
without a green blade of grass in the summer, and not a
drop of water for miles; immense swamps and morasses,
impenetrable even to the foot of the native, interspersed
with open lagoons and creeks, and water-holes, hidden
from the view by dense masses of tea-tree scrub; sandy
moors, clothed with coarse stunted heather; the distant
horizon, bounded by heavily-timbered ranges, extending
throughout the country, form the principal features of
the Australian landscape on the shores of Port Phillip,

where nature still reigns paramount in her sternest and
wildest mood.

Scarcely a wild-flower of any size delights the eye,
except the pink and white heather, which certainly do
present a splendid appearance when a large patch in full
blossom bursts suddenly upon the view; and some creepers,
which, beautiful as they are when examined closely, are
too small to attract the notice of the casual passer-by.
The wild orchis and the geranium are everywhere common : many of the smaller species appear to be identical
with their namesakes at home; and I have plucked more
than one little wild mountain-flower on the ranges here,
which has brought back to my mind scenes many thousand miles distant. The flowering shrubs are some of
them most beautiful; and many a rare exotic, which
would be highly prized in the greenhouses at home,
blooms unnoticed in all the wild luxuriance of untamed
nature, in the deep gullies here. The only two large
wild-flowers that I ever saw, were the large white lotus,
common in the water-holes and creeks, and a large white
star-flower, which grew in rich profusion by many of the
mountain streams. As for wild fruits, I never could fall
in with any worth gathering, except the wild cherry,
which is a little larger than an apple-pip, and grows
with the stone outside; the native grape, of a transparent greenish-yellow hue, as large as a black currant,
which clusters on a creeper thickly interlaced among the
tea-tree scrub; the cranberry, which here grows on a flat
creeping bush, on and sometimes in the sand; and a few

other berries no larger than currants. The Australian cranberry, which is described as growing on a bush ten to fifteen feet high, and the berries of which resemble the Siberian crab, does not grow here. There is a fine fruit peculiar to the "mallee" scrub, of a bright red colour, called the "quontong," about the size of a greengage, which grows on a shrub something like a small shey-oak; it is bitter to the taste, makes excellent preserves, and the emus eat them greedily. I have tasted a kind of fruit they call the native pear, not half so good to eat as a raw Swedish turnip. Whatever others may be found in the interior I do not know; but, as far as regards handsome or remarkable species, both in botany as well as ornithology, this district must be about the worst in Australia. Melons and pumpkins will grow anywhere, if planted, wild; and a delicious fruit is the little watermelon in hot weather.

There is a savage grandeur in the scenery of the Victorian forests, unsoftened by the lighter foliage of those beautiful shrubs which we generally look for in a southern land. The principal features of the woodland landscape here are old gum and huge iron or stringy bark trees, which have braved the storms of centuries, and stand out in bold relief from the deep evergreen of the cherry, the light foliage of the wattle or blackwood trees, and the "mournful weird-like appendages" of the shey-oak. The gullies are choked with shrubs, many of them very beautiful; but we see little variety among the forest trees; and we look in vain for the

graceful pine and silvery birch, which add so much to
the beauty of the northern forests.   The ash, the beech,
and the elm, which give so softened an appearance to the
woodlands of England, and the leafy palm, the stately
date, and other light feathery trees that grace the
tropical landscape, are strangers here.

As Prichard observes, " Here we do not find in the
great masses of vegetation either the majesty of the
virgin forests of America, the variety and elegance of
those of Asia, or the delicacy and freshness of the woods
of our temperate climate of Europe."

The vegetation is gloomy and sad.   It has the aspect
of our evergreens or heathers.   The plants are for the
most part woody; the leaves of nearly all are linear,
lanceolated, small, coriaceous, and spinescent.   The
grasses, which elsewhere are generally soft and flexible,
participate in the stiffness of the other vegetables.   The
greater part of the plants of New Holland belong to new
genera, and those included in the genera already known
are of new species.   The native families which prevail
are those of the heaths, the protæ compositæ, leguminosæ,
and myrthideæ; the larger trees all belong to the last
family.

Some of the trees, however, are very pretty, especially
the cherry, the box, and the wattle.   The cherry rather
reminded me of the yew at home, with its dark sombre
foliage; and I have at times fancied a resemblance
between some of the old gum-trees, with their gnarled
and twisted branches, and the brave old oak of happy

England. The forests here are open, and there is little undergrowth in many places. But the ground is everywhere covered with dead logs and branches, which lie rotting in the wind and sun—a sure sign that the hand of man has as yet but little interfered with the works of nature. It is curious that most of the trees here are rotten at the core; the wood of many is very brittle, and huge limbs are continually splitting and falling from the trees without any apparent cause, and with but little warning to the passer below. The wood of all is heavy, sinks in water, and splinters much; and although, when polished, the grain of many of the trees is beautiful, the wood is in general too hard and stringy to be used much for domestic purposes. Posts and rails, and large slabs for building bush-huts, are about the only uses to which the Australian timber can be put. If this country had only the pine forests and rivers of northern Europe, it would be perfect. But Nature divides her gifts, and what she denies to one land, she bestows on another. If the pine and fir could be introduced into these forests, what a boon would be conferred upon the inhabitants of Australia in the next century! The kauri pine of New Zealand might surely be grown here. All British trees and shrubs thrive in Australia, and in many a settler's garden we see standard-peaches, nectarines, and rose-trees flourishing, without any artificial aid, in a climate that here renders those trees and shrubs common to all, which at home are but the property of few.

I have seen some fine gun-stocks made both of cherry

and light wood, but although they may be prized as curiosities, for use they are not to be compared to a good bit of walnut. As a hint, however, I may here add, that if a man fancies a tree for a gun-stock—and many of the cherries and light-wood trees, which are the best for this purpose, have a capital curve—he should cut down a tree, not too large, hack it roughly into shape, and bury it in the ground for some time to season, and it will not split.

About thirty-five different species of trees, whose wood can be put to useful purposes, are, I believe, known to grow in Victoria.

The gum is certainly the Australian oak, and monarch of the forests here; but although it often grows to an immense size, and is a handsome wide-spreading tree, it has not the majestic or durable appearance of the British oak: moreover, the wood is worth nothing in comparison; the butt is very short, for the branches begin to spring out at a short distance up, and the wood is hard and splintery, and little good, except for firing. The branches are more crooked and twisted than in any tree I ever saw. The trees soon rot; and an old gum-tree shows more signs of premature decay, than as being fairly worn out by old age. The bark is as smooth as glass, and the trees very difficult to climb. The leaves are long, and these grow in drooping clusters, of a pale-green colour, white under. The blossom is pretty, of a pale-yellow colour, in thick bunches, and the pollen and honey from them afford a rich treat to the honey-eating birds of these forests. There are many varieties of gum; the leaves of one of which—the peppermint—have a strong peppermint flavour, to which the opossums are very partial.

The stringy and iron bark trees also—I believe, species of gum—are commoner than the gum in certain places. In fact, each tree seems to fancy a peculiar locality, and rarely grow together. I remarked that the gum-trees grow in much moister situations than the others. These trees grow much straighter, cut out into greater lengths, and the timber of these and the "messmet," as we called it—a species of bastard gum—were much more used in our forest than the gum, especially for posts and rails.

The bark, when properly stripped off, is very useful for thatching bush-huts, flooring tents, &c.; and the coat of inner fibres inside the bark might be put to many useful purposes. The Blacks also make canoes of it; but the only native canoes that I have seen have been "dug-outs," similar to those used by the North American Indians; but the Blacks in our district required no canoes.

Splitting posts and rails is a good bush trade, when men understand which trees split well, and pick a country where good trees stand thick. About 25s. is the general price per 100 on the ground: and two men will knock out more than that in a day. All the capital required to start this trade is a crown license, tent, tools, and rations, and a strong arm. And very different is the life of the Australian wood-splitter to that of the

North American lumberer. Here the wood-cutter is always dry, and his life is a healthy one; whereas premature old age and shortness of days are the inevitable fate of the timber-lumberer on the banks of the American rivers. Cutting firewood along the coast used to give employment to many. But, I believe, no timber must now be cut within a mile of the beach; and it won't pay for cutting at any distance, unless there is water carriage to Melbourne. The wood must soon become very scarce around the towns.

The wattle is a pretty, light-looking, flowering shrub, in some places growing to good-sized trees, but generally tall shrubs; and the wattle-poles are used for a variety of purposes. The bark is much sought after for tanning; and gathering wattle-bark, in places where the trees stood thick, paid well at one time. As the tree dies directly the bark is stripped, we rarely now see a wattle-tree of any size in the forests where the bark-peelers have been at work.

The honeysuckle is certainly the most worthless tree in the forest, and I think one of the ugliest. The wood is spongy and soon rots; there is nothing handsome either in the shape or foliage of the tree, the branches being crooked and ragged, and one half of them generally rotten. The wood is no good, even for burning, as it holds no heat. In autumn the trees are covered with large cones, which, when flowering, are clothed with a yellow down, filled with sweet pollen, and on this account the honey-eating birds much frequent the honeysuckles.

Here and there they grow to an immense size, but you generally see them ten to twelve feet high, growing together in patches in the gullies and damp ground. Inside the dead cone is a kind of pith, which makes a famous wick for a " slush-lamp." It often happened that we were out of candles in the bush, and we could not run into the grocer's over the way, and buy a pound just when we wanted; so we filled an old panikin half full of sand, stuck in one of these small honeysuckle cones, melted a little fat, poured it on to the sand to fill the panikin; lit the cone when it was hard, and this we called a slush-lamp.

The shey-oak is a prettier tree, but never grows to any great size, has a wide-spreading top, and the leaves are peculiar, being nothing more than long drooping fibres. The shey-oak apple is a pretty kind of fruit, resembling the cone of some of the species of pine, but small and round; it is bitter to the taste, and used in flavouring the colonial beer. The wood is hard, makes capital fires, and the root of the shey-oak is much used by the Blacks in making their weapons, such as bomerangs, liangels, &c.

The light or black wood tree is another pretty tree, rarely grows to any large size, and the wood, for beauty of grain and general utility, beats any in these forests. The blossom is pale yellow, and has a beautiful scent, much like that of the lilac.

The swamp-oak is an elegant shrub, and put me much in mind of the broom at home. The flower is yellow, and

Q

when in full bloom it has a very pretty appearance, waving
in the light summer breeze. They generally grow in patches,
on moist ground, and in the old country would be the
very cover to hold a spring fox or an outlying pheasant.

The tea-tree is the common scrub here, and grows
universally throughout the country, in dense patches in
all moist situations, and by the side of creeks and water-
holes; and it is to the cover of this shrub that the
Australian shooter owes many a pair of ducks. It is a
kind of high bush, grows on a long pole up to twelve to
fifteen feet; the branches are bushy, shoot up always per-
pendicularly, and the leaves are something like the spines
on a fir. It has a very pretty yellow blossom. The tea-
tree grows of all heights; is of a dark-green colour:
the scrub is very thick, and almost impenetrable, except
down a cattle track. The wood is hard; the poles are
straight, and valuable for many bush purposes.

A species of mistletoe grows as a parasite on some of
the old gums. It strongly reminds one of the mistletoe
at home.

Immense pieces of swamp, or fungus, are found in the
forest, and, when dry, used to make an excellent tinder
for lighting our pipes. I have seen the Blacks eating a
kind of fungus; and there are several edible roots here
which we know not, but which they grub up and eat.

In the summer, a kind of white secretion gathers on
the leaves of some of the gum trees, and falls to the
ground. It is here called manna. It is sweet, and not
unlike coarse pounded sugar. I never saw it in such

quantities as would pay for gathering, unless it was very clear.

The general scrub or underwood of these forests is "myall;" some species of coarse heather and low shrubs. I do not know of any thorn in this country except the box.

I have above slightly noticed the common trees and bushes peculiar to this district. Of course, there are others which I do not know, but these are the most striking ones. The fern tree is met with in the Dandenong ranges; but many handsome and curious species, such as the cedar, the pine, the cabbage-palm, and the grass-tree, peculiar to other parts of Australia, are strangers here.

Nothing, perhaps, in this country, strikes the English emigrant so much as the perennial greenness of the forests, and the slight difference there is in Australia between the summer and winter landscape. It is true that, in the winter, the leaves want the green freshness of spring, but they still cling to the branches; and at this season the trees shed their bark, which hangs down in long strips, waving to and fro as the wintry wind whistles through the forests with a low mournful wail. The seasons, moreover, so imperceptibly glide into each other, that the change is scarcely noticed. Magnificent as are the Australian forests, he cannot help at times comparing the monotony of the woodland landscape here, with the varied changes of scene which the different seasons present in his northern home; and at no season is the contrast so apparent as in the winter. "Enter a forest," as Inglis prettily observes, "when the sun breaks from the mists of morning upon the dews of the past night. Beautiful as is a forest in the spring, when the trees unfold their virgin blossoms; beautiful as it is in the summer, when the wandering sunbeams, falling through the foliage, chequer the mossy carpet beneath; beautiful as it is in autumn, when the painted leaves hang frail; it is more beautiful still when the tall pines and gnarled oaks stand in the deep silence of a wintry noon, their long arms and fantastic branches heaped with the feathery burden which has never caught one stain of earth." Such is a true picture of a northern forest, which the Australian native has never witnessed; and although probably out of place here, it may serve to remind more than one reader of many a distant scene, which, however rude in comparison to the soft and sunny landscape of the south, can never be fairly obliterated from the mind of that man who, let him wander where he will, still looks upon the land of his birth as his own peculiar home.

Notwithstanding its changeability, the climate of Victoria must be as salubrious as in any part of the world, or we could never lead the gipsy out-door life we do here: in fact, as a modern writer truly says, "the heat brings no fevers, the rain no agues, the cold no consumptions: the rivers are not bordered by miasma; the plains are bracing; the air pure; the sky open, blue, and bright. The bush itself is free from forest poison. The settler can range the land day and night, over hills, downs, prairies, and bush, sleeping in waggons or on the sward, without any fear of malaria to blight the healthy, or insidious fogs to undermine the delicate."

As to sleeping in the open air, except just in the rainy season, the bushman thinks nothing about it, if he has but a few matches and tobacco (and what bushman is without these)? But lighting a roaring fire, he throws himself down before it, with a saddle or game-bag for his pillow, and tumbles off to sleep as sound as in a favourite "four-poster." Night after night have I come in wet through from flight-shooting, and thrown myself down in my wet clothes on the tent floor for a few hours' nap till it was time to start for the morning's shot; and as for my old mate, he was a perfect water-dog, and when he came in wet I never knew him change his clothes, but he just sat before the fire till he was dry. All constitutions are certainly not alike; but I can only say that I never had a day's illness in the colony except it was brought on by my own imprudence, and I am certain no one led a harder or more exposed life.

The real Australian summer and winter are of short duration; the spring and autumn long, and the pleasantest seasons of the year. There is little or no twilight in the evening, and night sets in as soon as the sun sinks into the west—one deep crimson streak

> Of intense glory in the horizon's brim,
> While night o'er all around hangs dark and dim.

On June 21, which is the shortest day here, the sun
rises about 7, and sets a little after 5 P.M.; and on
December 21, the longest day, it rises a little after
4 A.M., and goes down about half-past 8 P.M.   The sun-
sets are sometimes magnificent.   The moonlight nights
are very beautiful; and at times the heavens present a
splendid appearance, spangled with myriads of stars,
many of them strangers to our northern hemisphere.
My favourite constellations were Orion's Belt and the
Southern Cross, for they have been my guiding stars on
many a night when the shades of evening closed in over
the forest, and found me miles away from my bush-tent.
A man soon becomes accustomed to find his way about
the bush.   A little knowledge of the proper position of
the sun, moon, and stars, and a small pocket-compass,
the bushman's surest friend, will guide him day or night
through the most trackless forests.   If ever the night
comes on thick, the best plan is to " camp up" at once,
for when the track is fairly lost, the more a man wanders
about to find it, the more confused will he become.   I
never needed a watch in the bush.   The sun is always
due north at twelve, and the reflection of its rays on a
compass will show the time within about a quarter of an
hour.   The hottest months in the year are January and
February, and then at times the heat is intense.   The
greatest curse to this country are the hot winds and bush-
fires.   In the summer evenings the sun sets fiery red,
and the bushman well knows what this betokens.   Next
morning the burning north wind comes sweeping over

the deserts of the interior like a furnace blast, and both
man and beast feel its effects. All bush-work is sus-
pended during the heat of the day. The cattle seek the
nearest water-holes, the dogs lie panting in the shade,
and the birds sit listless on the branches with open
mouths and drooping wings. The hot wind generally
commences about nine, and in the afternoon it suddenly
chops round to the south, and a sea-breeze sets in. The
hot winds are most prevalent in January and February;
they sometimes last three days, and come on perhaps
every ten days. In Melbourne a hot-wind day is called
a "brick-fielder," on account of the dust, which darkens
the sky. On these days dense volumes of smoke rising
in different directions warn the settler that bush-fires
are raging, and the whole country will be in a blaze perhaps for miles. The fire comes rolling on, devouring
everything in its progress, sweeping through forest and
over plain, and nothing stops it except a bare place that
has been previously burnt. Such a day was Black
Thursday some few years since, when bush-fires raged
throughout the country, and the loss of life and destruction of property was so immense, that it has ever since
been a " black-letter day" in the bushman's calendar.
In my opinion these fires are in a great measure pro-
duced by the heat of the sun and atmosphere. They are
sure to rage on a hot-wind day, and I have seen a fire
break out in different places on the plains a mile apart
at the same time. Glass bottles and pieces of tin lying
about act, I fancy, as burning glasses to the sun's rays,

and there are plenty of these all over the bush; for wherever the white man camps, he is sure to leave behind him an empty bottle or sardine case. The shooter should be very careful about his wadding in the dry summer-season. The law is stringent with regard to camp-fires in the bush during the summer. In pitching a tent in the forest, it is best to pick an open place in the timber, burn away the long grass all round, and clear up the dry logs. We were once burnt out, "lock, stock, and barrel," and lost all our little property.

Nothing can exceed the pure salubrity of the climate here in the spring and autumn, and there is a freshness in the early summer morning's breeze, laden with the perfume of the gum, lightwood, and other blossoms, which I never felt elsewhere. The nights are coldish throughout the year, and a heavy dew often falls in the end of summer. I never saw snow in Victoria, although I believe the snowy ranges and Australian alps are often covered. I only once saw ice about the thickness of a sixpence. Hailstorms are prevalent in the summer, often accompanied with thunder, and the hail-stones are about the size of marbles. The rainy season is very irregular. In some years the swamps and water-holes are well filled by July; in others the heavy rains do not fall until the end of winter. May, June, July, and August, are generally the rainy months. The drought is most felt in January and February. In most of the other months we have occasionally a rainy night or two. The principal weight of rain falls at night; and when it does rain, it comes down in earnest. It, however, soon runs off the dry rises, and sinks into the light sandy soil; and we suffered little or no inconvenience in the forests from the rain, even in the most rainy seasons.

I only saw three or four regular thunder-storms in this country; but these *were* storms, and one of them I shall recollect to the day of my death, for I was lost in the forest that night, and the storm raged with unabated violence from ten at night to three in the morning. The night was pitchy dark, except just when a flash of bright lightning lighted up the gloom of the forest, and then for some seconds I was fairly blinded. The rain came down in torrents; and as the blue lightning hissed through the sky, every flash seemed pointed where I stood, and a large shey-oak was struck, and shivered into a thousand pieces within a few yards of me. I never expected to leave the forest alive, and I think I never spent so long and dreary a night as this. I have looked death in the face more than once, but it has always been in moments of excitement. In this instance I stood helpless as a child, without any power to avert or even avoid the danger. Never does man feel his own insignificance so much as when the elements are at war around him. There is often a great deal of beautiful sheet lightning at night, especially in the northern sky, with distant rolls of thunder; but heavy thunder-storms are of very rare occurrence. I once felt a slight shock of an earthquake here.

I have perhaps already dwelt too long upon this

subject, but I must beg of the reader to bear with me a
little longer, and read a few lines which I have copied
out of *Melbourne Punch*, and which, although apparently
the offspring of a discontented mind, convey a true and
not very exaggerated idea of the natural history of this
land of contrarieties. If intended as a sarcasm, they
certainly contain more truth than bitterness.

Know ye the land where the shey-oak and gum-trees,
In shapeless deformity darken the wold ;
Where the blast of the north, where the chill of the sea breeze,
Now scorches to fever, now pierces with cold ?
Know ye the land contrariety sways,
Perverting the laws common nature obeys ;
Where black swans and magpies in whitened array,
And water-rats duck-billed, come forth to the day,
Where trees shed their barks as the serpents their skin,
And the stones of the cherries are outside, not in ;
Where the crowing of cocks at the midnight is heard,
And beasts breed their young in a manner absurd ;
Where enjoyment a fiction is, comfort a myth,
And the heart of an esculent hardens to pith ;
Where a wooden pear offers the toughest of fruit,
And the laugh of the bush jackass never is mute ;
Where the dust of the earth, and the glare of the sky,
Are a plague to the breath, to the skin, and the eye,
Where waters are brackish, and rivers are dry :
Where the load-star of life is the gold in the mine,
And the spirit supreme is the spirit of wine ?

CHAPTER XIV.

ALTHOUGH the *auri sacra fames* is the ruling passion of the colonist, there is a decided taste for field-sports, in fact, for manly sports of every description, among all classes out here. No one enjoys a race, or a good day's shooting, more than the Melbourne citizen; and a man of business is far more independent here than at home, for he can have an occasional run in the bush, and return again to his office, or counting-house, after having blown the Melbourne dust off him, without being "spotted" as a sporting-doctor, or lawyer, as the case may be. I certainly never, even among the game-keepers of the old country—and I had some experience with these gentlemen in my early days—have met with better field-shots, take them as a class, than the shooters out here; and as for riding, half the bushmen live in the saddle; and it would puzzle some of our best cross-country jocks to stay a couple of miles with three or four stock-riders, cutting out a wild bullock from a mob in the bush.

It is not astonishing that the turf is rapidly rising in Victoria, where horses are in such general request, and where the encouragement of a good breed is a matter of such vitality. A better and different breed of horses is gradually creeping into the country. Importations of thorough-breds from England yearly take place; more attention is paid to breeding and training; race meetings are springing up in every district; lads from most of the English stables find their way over here; the stakes are well worth winning; and now the impetus has once fairly been given, we may expect to see Victoria soon rank second only to England in the noble sport of horse-racing. There is no want of either money or pluck among the Victorian turfites; and when men of judgment are backed by men of money things are sure to go on right. I believe there are now more races in Victoria throughout the year than in any other country out of England. It is true there is a good deal of "leather-plating," and many of the colonial cracks have more the appearance of good English cocktails than "Derby winners;" but Rome was not built in a day, and if old Frank Buckle could rise from his grave, he would see many an alteration and improvement in turf matters at home which in his time were never dreamt of. Melbourne Course lies perhaps three miles out of town, and is as nice a country race-course as a man would wish to see. There are two separate clubs in Melbourne,—the Melbourne Jockey-Club and the Melbourne Turf-Club. Each has its separate meetings. The spring Jockey-Club meeting of 1858 lasted four days, during which twenty-one races were run. The public money given amounted to about

£2,000, and the gate-money came to near £1,000.
There is also a "convincing" ground on Emerald Hill,
near Melbourne, where private matches and steeple-
chases come off, and where many an owner is *convinced*,
to his cost, that his nag is not the flyer he took him to
be. A very fair new Government race-course has been
lately laid out, at Dandenong, about twenty miles from
Melbourne, where the managers get up two days' sport.
These races are principally supported by settlers in the
neighbourhood, and will see a better day. Independent
of the local interest they possess, they will afford a kind
of "hay and straw" meeting to the Melbourne trainers,
and occasionally put an odd £50 into the pockets of
those men whose stud is limited, and these hardly able
to run in the best company. Within the last two years
a new era has opened upon the Victorian turf, by the
establishment of Mr. Anthony Green's training-stables,
on Emerald Hill. He certainly led off the ball well, for in
the year 1857, his first year of public training, his stable
won sixty-one races; value of the stakes, £8,693. No bad
work this for a young colony. If a knowledge of his
profession and experience are worth anything, he ought
to get on. One of the oldest and best friends I ever
had in the colony, he has my sincere wishes for his
success. I trust he may long keep up the position he
now holds, as the " John Scott " of Victoria; and may
" Green's lot " prove a terror to the Victorian turf for
many a year to come!

A small but remarkably neat pack of hounds is kept

in Melbourne, hunted by Mr. George Watson, of steeple-chase celebrity. Their meets are advertised, and their doings duly chronicled in *Bell's Life in Victoria*, by " Nimrod's Ghost." As they rarely meet more than ten or twelve miles from the kennel, and the wild-dog and kangaroo are now rare in this district, a " bagman " is generally the order of the day; and I rather fancy that the meets of the Melbourne hunt bring out men more to try the merits of rival nags and riders among the stiff three-rail enclosures of this country, than to show the real science of hunting.

It is little wonder that steeple-chasing should be a favourite amusement in this land of " posts and rails;" and Abbott and Walkover will long live in the recollection of all who have seen this game little nag and rider leading the van in a good field of Victorian steeple-chase cracks. Trotting-matches and foot-races are constantly taking place. Regattas are often got up in the bay. There are some excellent wrestlers on the diggings. Border gatherings, periodically held in different districts, recall the memory of the Scot to the games of " Auld Lang Syne;" and every Christmas some sporting and eccentric " Pub." advertises amusement for the million, in the shape of sack-races, climbing the greasy-pole, and other good old English sports.

Pigeon-shooting from the trap seems to be on the rise, and every season has a champion. Just as I was leaving, two or three heavy matches were on the carpet.

My old friend, Goldstone, was at the top of the tree; and although, no doubt, there are as good fish in the sea as ever came out of it, yet when he walks up to the trap, I should consider it a safe investment to " put a fiver " on him.

" The ring " is not forgotten in Victoria; and in a new country like this, where men of all nations are thrown together, where the bowie-knife and the revolver are weapons familiar to all, where two out of every three men we meet are in the prime of youth and strength, where a sort of Jack's-as-good-as-his-master feeling is predominant among all classes, where in a hasty dispute a blow quickly follows a word—the encouragement of fair British boxing is of far greater moment than might at first be supposed. For of all modes of settling a dispute, the naked fist must be the best, so long as the rules of the British P.R. are fairly acted up to. One thing is certain, that although we occasionally do hear of the most dastardly crimes being committed, the general feeling of man towards man in this country is good; and if his quarrel is but just, a man may generally reckon on having fair play. There are several professors of the " manly art of self-defence," men or stars of the London ring, both in Melbourne and on the diggings, and plenty of the right stuff to make good boxers. *Bell's Life in Victoria* is the oracle of the Victorian P.R., and when a " tournament " does come off, all things are conducted with as much order and regularity as if the veteran commissary were there with his staff, and the veritable

"Nunquam Dormio" watching the proceedings with his eagle eye.

Bell's Life in Victoria, a new periodical, devoted to "the turf, the chase, and the ring," is published weekly, and in addition to all the sporting news of the colony, generally copies from its London namesake a *verbatim* account of any good fight or race. The time for news to travel between home and here is short, and when any great sporting event is about to come off in the old country, it is looked forward to with as much interest here as at home. *Bell's Life in Victoria* appears to be ably conducted, free from all party spirit, and such an organ must be of the greatest service to the sporting interests of the colony.

Of all out-door manly sports, however, cricket has gone ahead more in a short time than any game in the colony, and Victoria can now pick an eleven which will be soon hard to beat by any club in the world. It is plain there must be some very good professionals out here as tutors, for I have watched practice on the Melbourne ground which would not disgrace Lord's, and the club is but yet in its infancy. But the Victorian cricketers seem to enter heart and soul into their favourite game, and stick at nothing to render their club perfect. This is as it should be, and it is gratifying to see that those manly old English games, which were the pride of our forefathers, and have rendered England famous throughout the world, are not forgotten when her sons leave their native shores, but meet in friendly rivalry in foreign climes.   Matches are played between the Victorians and the Sydney club; and although I believe at present the old colony has the pull, we may expect, as practice makes perfect, to see the tables turned, and "the old man beaten by the boy."

I have seen some capital black-breasted Reds out here; but if any cock-fighting is carried on, it is done "under the rose."   And although there are some most "varmint" looking "tykes," both in Melbourne and on the diggings, they appear to be kept for their legitimate purpose—that of guarding the house or tent.

There is a Tattersall's-yard in Melbourne, with an hotel attached, where turf business is transacted.   There is the Turf-Club Hotel, several private clubs, and a public betting-room in Bourke Street, where a man can "get pepper" to any amount about what horses he fancies best.   Business is also done here on the home events.   Several horse-bazaars are held in the town every morning, where the "horsey" gentlemen of the neighbourhood are wont to congregate and compare notes; and if a sporting man is wanted, he is generally to be found in Row's Bazaar, or Watson's Sale-yard.   There are public billiard-rooms for the nobs, skittle-alleys for the mob, and chess divans for the sober coffee-drinkers; and each class is ably represented.   Much to the credit of the town, there are no public gaming-houses, although I have not the least doubt that, if a man fancies himself, he can anywhere, "on the quiet," find his match at a hand of crib or blind all-fours.

R

Take it all in all, Melbourne bids fair soon to become one of the most sporting towns out of England. There is nothing strait-laced about the colonist. If he wishes for a day's sport, he has it; and he backs his opinion in a race, fight, or steeple-chase without caring who knows it; and is thought none the worse of for it. Moreover, the tastes of most men out here are this way inclined; and as long as this is the case the good cause must flourish. Rational and manly amusements will always create a good feeling among the inhabitants of any land, and nothing shows the character of a people more than the choice of their sports and pastimes. The encouragement of field sports gives a healthy tone and manly bias to all classes, which will, I trust, long continue in Port Phillip; although I myself have ceased to be personally interested in the doings of the colony.

CHAPTER XV.

RIVER FISH AND ANGLING.

VERY little can be said respecting the angling, or river-fishing, in Victoria. Coming, as I did, into this country with the magnificent rivers and lakes of Sweden fresh in my mind, nothing struck me more than the insignificance of the rivers in this part of the colony; they are, in fact, little more than creeks—many of them merely a succession of water-holes, during the summer choked with high reeds and bulrushes; the banks of others stony, steep, and rugged, or grown up with small trees, which overhang the stream and nearly meet at the top. I brought all my salmon-tackle out with me from the north: the only use I made of my rod was to cut it up for cleaning and ramrods. My trolling and fly-lines came in handy for tying up wisps of game; and my salmon and trout flies soon became the prey of the moth, instead of the fish. A man requires very little fishing-tackle out here, and whatever he wants he can buy in Melbourne. Most of the creeks and water-holes, however, do abound with fish, such as they are; but the only fresh-water species I ever met with were eel, bream, trout, black-fish, mullet, and herring; and most of these can live in salt as well as in fresh water.

The Australian eel runs to a large size, is blacker in colour than the British eel, and fatter and richer to the taste. They abound in all the swamps when they are full of water, and as the water runs off they get into the creeks and water-holes, where they remain during the summer, and early in autumn they draw down to the sea. I cannot say whether they go there for the purpose of spawning, but I know these migrations are very regular, and thousands are caught about the end of March and beginning of April, at the mouths of the creeks, as they travel out. When they come back I don't know, but they do not appear to remain long in the sea, for I used to see the Blacks spearing them on the swamps as soon as ever they filled and the creeks began to run fresh-water. During the winter all the creeks that have outlets to the sea are fresh, and the water in the bay at their mouths is only brackish for a long way out to sea; whereas, in the summer, the water in these creeks is salt for a long way up. In many places the eels here don't care to take a bait, but where they do a large worm on a night line always answers. It is a singular fact that earth-worms are very rare in Australia. The Blacks are very expert at spearing eels on the swamps with a single long-pointed spear; I have known two of them to catch 1 cwt. in a day, and sell them for a bottle of grog. The wholesale price here is generally 3*d*. per lb. I never saw a real good eel-gleave out here— the very thing for these swamps and creeks.

The black-fish is found in most creeks, and appears to be

about the only fresh-water fish, except perhaps the eel and the trout, indigenous to the inland creeks, which have no direct communication with the sea. The black-fish runs to no great size, at least with us—I never saw one weigh 1 lb. In appearance and habits it rather resembles the tench at home, and is a nice-eating fish.

The Australian bream, or brim, was certainly the best of all our fresh-water species; but I am not certain whether it can be strictly called a fresh-water fish, for I have only seen them in the creeks during the summer, when the water was brackish. It is something like the European bream in appearance, but not so coarse, and more silvery to look at. With us they rarely exceeded 1 lb. in weight, and were very much like the small snapper. Its habits much resemble those of the carp—very shy, frequenting deep clear holes; and were I to angle for them I should fish early and late, and use the same bait and tackle as for that fish.

The Australian river trout is a very poor apology for the trout at home, and looks more like a cross between the roach and the miller's-thumb than anything else. I rarely saw one longer than my finger. It seems to be a grovelling fish, does not rise to the fly, and I fancy we were wrong in calling it the trout, which fish it resembles neither in shape, habits, or appearance. They were common in most creeks and water-holes.

At certain seasons the herring and mullet come up from the salt water into the creeks and rivers that run direct into the sea, and they appear to live as well in fresh as salt water. They are the only fish that will rise to the fly here, and catching them is about on a par with whipping for dace and roach at home. For fly-fishing the Yarra and the Barwen are the Coquet and the Dove of Victoria. I believe the general fly in use out here is red body and white wings; but the Victorian fly-fisher requires no very varied assortment of flies in his book. Both the herring and mullet are bright, clean-looking fish, but they run to no great size.

I believe the Murray cod is a fine fish; so large, according to the accounts I have heard, that I fear my readers would never be able to swallow them. I have, however, seen one over 20 lbs., and it resembled its European namesake in no one point that I could perceive. Cod-fishing in the Murray, however, is now becoming a lucrative trade. In the large rivers up the country, such as the Murray and Darling, there may be other fish and better fishing than on the coast, but my remarks apply only to the Melbourne district.

Small crayfish abound in all the swamps; and a small species of turtle is taken on the banks of some of the inland rivers, the eggs of which are considered a delicacy. There are no species of pike or perch in these lakes and rivers, and none of those soft fish, such as the roach, chub, or carp, peculiar to Britain. The flesh of Victorian fresh-water fish is certainly very meagre.

One thing is quite clear—that Victoria is no country for the angler. I hardly ever saw a stream on this side adapted to throwing the fly. Even if the fish were worth killing, and if you hooked a good fish, the chances would be very much against your landing it, owing to the steepness and rottenness of the banks, and the heaps of moss and rubbish with which every river is filled. Moreover, angling is never likely to become a favourite amusement with the present race of Victorians. Unless he can make good wages at it, neither the regular colonial shooter, nor fisherman, deems the sport worth following; and angling in this country hardly affords excitement enough for the amateur sportsman. Of all field-sports angling is, without doubt, the least mercenary, peculiarly the sport of youth and declining years, and a happy and contented mind. As long as the gold-fever rages there is not likely to be much quiet or content out here; no one in this country, as long as he can earn a shilling, is considered old enough to knock off work; and, as for the young " currency lads," they are more precocious than the youth at home, and cracking a stock-whip is more to their taste than throwing the fly.

CHAPTER XVI.

SEA-FISH AND SEA-FISHING.

But these coasts abound in sea-fish of many species; and sea-fishing, although, like many other things in the colony, now overdone, is still a paying game, when men are steady and stick to their work. Since the great influx of "Celestials," salt fish has risen in value; and if John Chinaman has benefited no one else in the colony, he has at least done some good to the fishermen; for instead of being obliged now, as formerly, to run the fish up to Melbourne themselves, or sell them to the hawkers at their own prices, on every fishing station along the coast Chinamen are camped, who buy the fish from the boat, and salt them on the spot. Tons of salt fish are yearly sent up to the diggings for consumption by the Chinamen.

There is always a ready sale for fresh fish in Melbourne, and often at exorbitant prices. A good fish-market is much wanted here. The Melbourne Billingsgate is held on Prince's Bridge, at daylight, where the hawkers from the country sell their fish to the street hawkers, and any one who wishes to hear a little chaff or colonial slang can enjoy a rich treat by paying a visit to the bridge-end any morning, when the fish-carts come up.

The principal sea-fish here are snapper, flathead, sea-pike, salmon, salmon-trout, mullet, herring, and gar-fish. There are doubtless other species, unknown to me, but these were the common market fish on our coasts. The snapper and flathead are about the only ones taken by the hook. There are some famous oyster-beds in Western Port Bay, and there must also be oysters in Port Phillip Bay, only no beds have yet been found, for I have often picked up capital oysters on this beach, washed ashore after a heavy gale. The wholesale price for oysters in Western Port Bay is 3*d*. per dozen. The oyster, peculiar to our bays, was large, and resembled the coarse British oyster in appearance and flavour; but there is a very pretty little shell-fish, which they call the Sydney oyster; this is not the shape of the common oyster, being long and deep, and the shell ridged. The pearl oyster, which is found on some of these coasts, is not met with in these bays. Some very large cray-fish are taken at the Heads. Shrimps used to abound in Port Phillip Bay; and when they sold at 3*s*. per quart, and a man could catch some gallons in the day, shrimping was as good as gold digging. An old friend of mine made a little fortune at it when thirsty gold diggers swarmed in the Melbourne public-houses. Strange to say, all the shrimps have now disappeared from this bay.

The snapper season is the fisherman's harvest here; they come on to these coasts about September (I recollect we used generally to kill our first snipe about the time the first snapper was caught), and are taken up to Christmas. They then leave the ground, travel up the bay by the Geelong line, and out to sea at the Heads. The snapper is a flat, coarse-looking fish, something similar to a large bream, but with a large prickly dorsal fin like the perch. They are a very gregarious, bold biting fish, have their particular feeding-grounds, and I have known six or eight dozen big snapper taken by one boat's crew in the day. They are worth about £1 to £1. 10s. per dozen, are a very good eating fish, and take salt well. The large snapper run from 12 to 20 lbs., and I have seen them larger. As soon as the big snapper begin to leave, the second-sized ones come on to the feeding-grounds, and last of all, about Christmas, the small ones.

The flathead is a curious-looking fish, and, like the morepoke among birds, seems all head. They generally run from half a pound up to two pounds, but I have seen them larger. They have a sharp prickle on the edge of each gill cover, the wound of which is dangerous. I have seen some very ugly wounds inflicted both by this and the prickly back fin of the snapper, and it is dangerous work standing in a crowded boat with the big snapper floundering about the bottom, unless a man has on good sea-boots. I suppose it is owing to the quantity of animal food men eat, and to the heat of the blood; but a small flesh wound, which at home would be treated as nothing, is often out here attended with serious consequences.

A poultice made of the leaves and stalk of the marsh mallow, which in many places here grows wild, and is the most valuable plant in the bushman's herbal, is an excellent remedy for cuts, bruises, swellings, &c.

The flat-fish come into the bay with the small snapper, when the large fish have left, and are caught up so late in autumn, principally with lines, while the boat is drifting. The small snapper are caught with bait on the same ground as the large ones; twenty to thirty dozen of flathead and small snapper are sometimes caught by two hands in the day. They will be worth about 2s. per dozen; but the value of the fish here depends much on circumstances. The flathead is considered as fine a table fish as any in the colony.

The principal net-fish here are salmon, salmon-trout, mullet, herring, seapike, and gar-fish; and these come on the coast, in large shoals, at irregular periods.

The salmon and the salmon-trout rather resemble the small salmon-trout at home in shape and appearance; but they have no adipose fin, and rarely run to any large size. They are both, however, clean, silvery-looking, nice eating fish. Now and again a good haul of salmon or herrings is made on the coast, and I have known a boat's crew to clear £60 in one night. But this is a rare occurrence now. The large shoals of fish don't set in to these shores as formerly; and if by chance one is seen, too many are on the look out to share the prey. As I said before, sea fishing is overdone in this bay. When I first knew Mordialloc, I don't believe more than three

fishing parties were camped there ; and " Wiseman," one
of the oldest and best fishermen in the bay, had the coast
nearly to himself.   Now there is a regular canvas town
of fishermen's tents here during the season, and I have
counted between forty and fifty boats on the snapper
ground at one time.

The scapike runs to 5 or 6 lbs., and much resembles
the scapike at home.   The gar-fish sometimes run to a
good size, are taken in large shoals, and sold by the
basketful.

There are a great many dog-fish in these bays, of
different species ; and one which we called the pig-headed
dog-fish is curious and interesting, as being antediluvian ;
in fact, many of the common fishes peculiar to these seas
are of the earliest kind, and have for the most part a
cartilaginous structure ; and it is worthy of remark, as
Professor Owen observes, that we have both in the
botany, zoology, and ichthyology of Australia a striking
analogy to that of the Oolite Æra (of geologists), a
period in the earliest stage of creation, when the mammalia first appeared.

I have seen some fair-sized sharks taken in both these
bays, and one monster, which must have rivalled Port-
Royal Tom, haunted our bay for a season, and if he were
only half as large as the fishermen represented, must
have indeed been a wonderful fish ; I don't think, however, we had any ground-sharks ; I never heard of a
whale finding its way in at the Heads ; at times heads of
large porpoises would show themselves, but neither in

numbers nor varieties of species can these coasts be at all compared to the British shores.

The benito sometimes, but rarely, comes into these bays. The butter-fish runs to a large size off the Heads; and if the accounts I have heard are true, this must be the largest eating-fish off these coasts. The smaller ones used to come on to our beach in summer, and we speared them in shallow water.

We had two species of large ray,—the one which we called the stingoree—for, I presume, the stingy ray; and the other the old maid, or fiddle-fish; and small flounders abounded on the sandy flats. The stingoree is a very large species of ray, often weighing 15 or 20 lbs., with a long thin tail, and a long, sharp, jagged spike on the back of the tail, which the fish can erect at pleasure. The fiddler is something similar, but rounder, with a smaller tail, and no spike. Both used to lie on the bottom in shallow water. The back of the fiddle-fish is marked with black lines, and I suppose it derives its name from some fancied resemblance to a fiddle. The livers of both these fish, as well as the shark and dog-fish, boil down to capital oil, and this is the only purpose they are put to, neither being considered eatable. I have, however, eaten both, and, with the help of a bottle of " Burgess's original," should not have known them from skait. Jelly-fish of all shapes and sizes float about the bay; and cuttle-fish, the long tendons of which are an excellent bait for snapper, and which we called squid, abound on the coasts. There are several nasty-looking fish in these bays;—the poisonous toad-fish, the prickly porcupine-fish, and others; and often, when a net is drawn ashore, many small but singular wonders of the deep are brought to light.

We had several species of limpet or wilks on the small rocks; one which we called "the warrener," as large as a great wood-snail, which was capital eating. By the way, I never recollect seeing any land-snails in these forests. But the finest shell-fish in this bay was the "mutton-fish," which in the island of Jersey is called the "ormer," a large flat shell-fish, often as large as one's hand, which sticks so closely to the rocks by the fleshy side that they require to be removed with a knife. These mutton-fish are excellent eating when roasted on the ashes, and a dozen of them will make what is colonially termed a "capital feed." The Blacks are very fond of them; and it is extraordinary to see what a time they can remain under water when diving for mutton-fish on the rocks below the surface.

Two or three species of small crab were found in the crevices of the rocks at low water; and one, which we called the soldier-crab, was handsome and curious. There was a funny little species of land-crab, round, and about as large as a musket-ball, which used to cover the beach at low water in such quantities in certain places, that the ground seemed alive with them as they scuttled backwards into the sand. The crayfish, however, represented our lobster and crab on these shores.

A small species of saw-fish—I have seen the saw about

one foot long—is met with in these bays; and there are some very pretty varieties of star-fish. But we had very few shells on these coasts: those which we did find were small and plain-looking; and whatever value they might bear in the eyes of a conchologist, were certainly not to be prized on account of their beauty.

Good fishing-gear is still dear out here, especially English sieve-nets and a good whale-boat: and to start right, a fishing-party requires some capital. I have known men stick to it during the summer, in a small dingy, single-handed, and make a good living, when the hook-fish were well in: but this is dangerous work; for the squalls come on so suddenly in these bays, that the fishermen have often to "up killoch" and run into shore before the wind with scarcely five minutes' warning. Four is about the right number for a good fishing-party; and if they only understood their business, worked steadily at it, and shunned the "nobbler"—the ruin of many a good man in this country,—they could hardly fail to do well. But, like the shooters, the fisherman's motto is generally "happy-go-lucky;" and perhaps the principal reason why we never see either in very flourishing circumstances is, that there is rarely a woman in the bush-tent to keep "the house in order."

The fur-seal abounds at certain seasons on some of the rocky islands at the entrance to Western Port Bay. The skins are valuable, and I should think the blubber was worth something: but nobody seemed to care much about them. Sealing, however, is not a boy's game; for it requires a good boat and hardy crew to weather the surf, which at times breaks with thundering violence over the iron-bound coast at the entrance to this bay. The seals appear to come on to these rocks about the end of November; and fine still weather in December is the right time for sealing.

Western Port Bay will, I fancy, soon be the great rendezvous of the fishermen south of Melbourne. The shipping and steamers have much disturbed the fish in Port Phillip Bay within the last few years, whereas in Western Port Bay there is no harbour for shipping; and although the shores are principally mud-flats instead of a sandy beach, there is much good fishing-ground, and many places where a net can be "shot." The distance from town is considerable; but even now hawkers run regularly during the winter; and depend upon it, if the fishermen once get down there, John Chinaman will soon follow them.

When I first came into this district, I camped for two seasons at Mordialloc, on the beach, about fifteen miles south of Melbourne, then the best fishing-station in this part of the country. In my time there was not a better shooting-ground anywhere near Melbourne; and had things only remained as they then were, I should never have cared to leave it; but the game became scarce, and all the land bought up, so that you could not walk a mile without a three-rail fence staring you in the face. I shall even look back with pleasure upon the time I spent at Mordialloc; nor shall I easily forget the uniform kind

treatment I received, not only at the hands of Mr.
McDonald, the owner of the station, but of every one
else connected with it. I certainly was more at home
there than in any other of my camping-places : for it is
very rarely that a station-master out here will conde-
scend to notice (otherwise than as a parish-beadle regards
a vagrant in the old country) a vagabond shooter who
camps upon his run.

A few years ago many a man earned good wages by
picking up "waifs and strays," washed ashore on the
coasts of this bay ; and I remember, when the *Ontario*
was wrecked on the Heads, in 1853, some thousands of
pounds' worth of property came ashore, and the beach
was strewed for days with valuable articles of every
description. Unfortunately, her cargo was not a dry
one ; and I saw a fatal accident, which resulted in the
death of one of our party, arise from the reckless manner in which the spirits were served out upon that occasion. Formerly every ship discharged her lumber in the
bay ; and owing to the heavy rates of storage in Melbourne, emigrants would cast many things overboard
rather than bring them ashore : now, however, people
are more careful ; and this beach is so regularly
"combed," that one rarely sees anything worth picking up.

CHAPTER XVII.

Of the Australian aborigines I have but little to say. They are a race fast passing away; and the few that we do meet with now about Melbourne—in fact, in all the settled districts—are very different men from the real Australian native of the last century. There are only two tribes now in the vicinity of Melbourne; and these are but remnants of what they were when we first took possession of their country. The Yarra Blacks, who camp about the ranges at the head of the Yarra, north-east of Melbourne, and the "Bomerang, or Coast Tribe," whose head station is at Mordialloc, and who own—if we can use that term now we have dispossessed them of all their land—the country to the southward down to the Heads. These, by constant intercourse with the white man, have learnt much of our language and habits, are on capital terms with us, and there is no more danger in meeting a lot of them in the bush than a gang of gipsies at home. The Gipps Land tribe appears to be the most numerous in this part of Port Phillip, and these men seem to be wilder and more ferocious than any I have seen. Wherever Government has taken up their land, a

Black's reserve of, I believe, a square mile, is left, and blankets and rations, provided by Government, are served out to them by the master of the station nearest to their reserve. There is also a protector, or kind of magistrate, appointed to look after their worldly interests; but no one seems to trouble himself about giving them any religious instruction. It is not within my province to offer any opinion as to whether or not it is our duty to do so, after, as it were, adopting them. There is a great cry at home about sending missionaries into foreign parts of which we know but little, and yet here we have tribes of savage heathens wandering about among Christians, in the close vicinity of a large city in a rising colony, which is now certainly more like England than any we possess, abounding in religious sects of all denominations, and yet no pains are taken to instruct or convert these poor savages. Perhaps it is not possible to do so. Perhaps they are better off as they are; and this is probably the case—for, as Bonwick justly observes, " we have a sad tale to tell when we speak of our so-called civilization upon these aborigines." To adopt our habits, they must be entirely removed from the associations of the Mia-Mia; and what have we to offer in exchange for endearing relations, joyous freedom, and an unanxious existence? The black man is thrust upon a competition society to earn his bread; he is exposed to the gibes and contempt of the lowest of our countrymen; he is without sympathy and without friends; and is herded with men from whom he learns the most obviously developed priu-

ciples of European civilization—swearing and drinking.
It is true he eats better food, wears better clothes, and
sleeps in better dwellings. But where is his home? Who
will be his sister, his mother, his brother? Who will
ally herself as wife to his dark skin? Can he ever know
the sweetness of a child's love? No! He soon tires of
our food, our work, our confined habitations, our heart-
less ridicule, and hastens back to his camp-fire, to find a
friend, to feel himself a man, to dwell with those that can
love him. Attempts were formerly made to convert them,
which always failed; but this was long before the country
was peopled as it is now. Heathens or no heathens, however, the life of the savage here is perhaps as free from
reproach as that of many of their Christian neighbours.

When I camped at Mordialloc, I lived on very neigh-
bourly terms with the "Bomerang" tribe, for they
generally had their "miamies" close to my hut; and
as I never made too free with them, or gave them a
promise I did not intend to keep, I was a bit of a
favourite with them. Like most other savages, they
strictly imitate the white man in all his vices; and this
tribe is fast paying the penalty; for since I knew it first,
more than two-thirds have been swept away by disease
and intemperance, and in a few years it will exist only
in name. It is melancholy to see a whole race of beings
thus disappear, without any apparent cause. There is
no prostration of physical strength, or mental activity;
they wither in the prime of life, and sink into the grave,
as though a blight had fallen on them.

Of the many thousands who inhabited this colony before the arrival of the white man, not 2,000 survive, and most of these are on the banks of the Murray. Although debased far below their own savage level since their intercourse with the white man, the few that are left still retain much of that free independent spirit, and wild roving disposition, which characterizes all savages who have to gain a living by the chase. For although they can get their rations all the year round at the head station, they never care to live long in one place; but, following up the habits of their early life, make periodical excursions into the bush at different seasons, when the different game is in. Thus swans' eggs, kangaroo, ducks, pigeons, eels, and crayfish, all furnish them with food and occupation at certain seasons; and it was but rarely that many of these were on the reserve at one time. I have often remarked, when wandering through these forests, that the Blacks invariably fix upon the prettiest situations for their camping places. I cannot help thinking that the character of the Australian aborigines has been much belied by those writers, who have described them as but one degree removed from the brute. It is true that they possess inherently all the bad qualities of the savage, and where is the wild man whose character is not marked by ferocity, treachery, or cunning? But they have also many good attributes, which might shame the white man. I always found them honest, and fond of the truth; and although they will ask for anything they fancy, just as if they had a right to it, I never knew them steal. They are a manly, independent race, certainly not cowards. Some of them are the merriest vagabonds under the sun. It would be impossible to make a slave of an Australian Black; and they always appeared to me to possess a degree of savage intelligence, superior to that of many other wild men. Some of the men are very athletic fellows, far from bad-looking; but I cannot say much for the personal appearance of the females. Strange to say, these ladies seem to care nothing for finery or ornaments, a dirty blanket, or opossum rug wrapped loosely round them, and a short black pipe stuck in their hair completes their toilette. The Black's opinion of the white man is pithy and laconic:—"Big one fool, white fellow, all same working bullock."

No improvements, or alterations, seem to surprise them. The Australian native, unlike his neighbour the New Zealander, makes no endeavour to keep pace with the times. "To be content, is his natural desire." The easier he can get his bread, the better he likes it; and if he can obtain sufficient food for the day, he cares little about the morrow. Nor is this to be wondered at, when he has been accustomed from his birth to lead a careless, wandering life, in a country where Nature has so liberally supplied him with food, and where the climate is such that a bush-gunny, ah, or mia-mia, will shelter him in the most inclement weather. Some of our chaps I used to like very much; and when my old friend, King Dermot, is gathered to his fathers, I trust his prediction to me upon one occasion will be verified—that " When he tumbled down, he should go up long way and fly about, all same big one eagle-hawk."

Although, as I have before stated, a fortune is not likely to be made by the gun out here, still I consider this is as much owing to the habits of the shooters themselves, as to anything else. If I were a second time thrown upon the shores of Australia, this is the life I should again follow; and if three good men—really working sportsmen, none of the make-believe sort—were to start with a small capital, fit themselves up with a house, tent, and rations, go down into some good kangaroo country near the coast, shoot and salt for the season, save the skins, as I have before recommended, and when the season was over, go down upon the beach and fish,—I am certain they might do as well at it as anything else in the bush. But the difficulty would be in finding three men who would stick well together for any length of time in this country, where self-interest is the only thing that binds men to each other, and where the whole decalogue appears to be comprised in this single sentence, " Man love thyself." Most men out here are red-hot to enter into any new scheme, but they will rarely stick long enough at it to give it a fair trial: as long as things go on right, all is well; but as soon as the sun becomes a little clouded, half of them knock under, and leave a mate without a moment's warning. It is strange, that although two men can, and do often, stick well together, we rarely see three agree long. Yet for this job there should be three, and if they would only give it a fair trial, they might make as good wages at it as any other bush work. I should, however, certainly not recommend either the labourer who can earn his steady £1 per week, or the man in town who has a regular and certain billet, to leave it and take to the gun. They are both better off where they are, and would probably be neither of them fit for this work; for it is a great mistake to suppose that shooting is a game to which any one may turn for an easy living when he can do nothing else. But for men like myself, who are neither labourers nor men of business, but who can at least handle a gun, and do not mind roughing it so long as they are free, this is the life; and I am certain that they would be far more independent, and I do not know whether they would not make as much at it as many a man in town, who, to all appearance, holds a good and lucrative situation. For although the profits may not be great, the expenses are small; and if it was not for " the bursts," which are almost sure to occur when a bushman visits town with the hard earnings of perhaps a twelvemonth in his pocket, he might always save a little money.

And now, in conclusion, a bit of advice to any old bush friend, who may chance to cast his eye over these pages. Unless his circumstances are such that he can live independent, or has good friends able and willing to help him, let him stay where he is, and not think of returning home. We all know what home sickness is; and where is the wanderer in a foreign clime, let his condition be what it may, who has not at times felt a longing desire once again to see the land of his birth? But old Time works his revolutions as steadily at home as abroad; and when he does return after a few years' absence, he will most probably find so many changes, so many ties will have been severed, that bound him to the home of his youth, so many old friends dead, others so changed, that in nine cases out of ten he will feel himself an alien in the land of his fathers, and experience far more regret than pleasure when he once again sets his foot on his native shore. Such was my fate; I trust it may not be the lot of all. Of one thing, however, I am certain, that the man who has led a wild bush-life for any length of time will hardly ever again settle down to the staid customs and formalities of the old country. He will miss the jolly freedom and independence of the bush; and this is the reason why so many who do go home with the intention of remaining, are sure to return again to the colony after a short absence. But, above all things, let no working man think of going home unless he " has made his pill;" for if he has to get his living by hard work, he will find it far easier to do so abroad than at home; and if he should chance to want a supper or a bed on this side of the equator, he will have something more to do to get it than to walk up to the first bush hut or tent that he comes across, and throwing down " his swag," by the simple passwords " Good evening, mates," obtain a hearty welcome for the night.

Cease we our chronicles, and now we pause,
Though not for want of matter; but 'tis time.

It is now some years since I left my home "a vagabond
to be," and during that period have wandered over many
lands, my gun and fishing-rod my only companions—a
true citizen of the world.

In the prime of years, in the full flush of youth and
strength, such a life offers charms of wild independence,
which can never be realized by that man who is tied to
one spot; no matter with what comforts he may be
surrounded, or what sport he may enjoy, ready made to
hand. But as years creep on, and a man begins to feel
that "the old gentleman with the scythe" is pressing
hard upon his heels, his enthusiasm will in a measure
abate; and the more he has buffeted with the rude waves
of the world, the greater will be his desire to cast anchor
in some quiet haven, which he may regard as a permanent
home in declining years. For how truly has Sam Slick
described the dark side of the wanderer's life in the following words: "Here to-day, gone to-morrow; to know
folks but to forget them; to love folks but to part with
them; to come without pleasure, to go without pain;
and at last, for a last will come to every story, still no
home." Never, perhaps, was the history of a life written
in so short a sentence.

Sterne wisely remarks: "Matter grows under our
hands; let no man say, come, I will write a duodecimo."
This must be my excuse if my wanderings have led the
reader too far. My fitness for the task I have under-

taken, I ground upon the fact of having lived five years in the bush, my sole occupation shooting and fishing; and as I have stated very little from hearsay, but nearly all from actual observation, the truth of all I have stated may be relied on. I have no intention of instructing the colonial sportsman,—my only wish is to amuse the sportsman at home. If this object is attained, and if every one who opens this little work feels half the pleasure in its perusal that I have done in writing it, I am satisfied. With this hope, and with best wishes for the welfare of all the old mates and friends I left behind me in the bush—and I did not know there were so many till I had to take leave of them—I shall close my slight notices of the field sports and *fauna* of Australia Felix.

INDEX.

COX AND WYMAN, PRINTERS, GREAT QUEEN STREET, LONDON.